Intelligent Systems Reference Library

Volume 80

Series editors

Janusz Kacprzyk, Polish Academy of Sciences, Warsaw, Poland
e-mail: kacprzyk@ibspan.waw.pl

Lakhmi C. Jain, University of Canberra, Canberra, Australia, and
University of South Australia, Adelaide, Australia
e-mail: Lakhmi.Jain@unisa.edu.au

Jean-Marc Mercantini · Colette Faucher

Editors

Risk and Cognition

 Springer

Editors
Jean-Marc Mercantini
Avenue Escadrille Normandie-Niemen
Aix-Marseille University
Marseille Cedex
France

Colette Faucher
Avenue Escadrille Normandie-Niemen
Aix-Marseille University
Marseille Cedex
France

ISSN 1868-4394   ISSN 1868-4408   (electronic)
Intelligent Systems Reference Library
ISBN 978-3-662-50921-0   ISBN 978-3-662-45704-7   (eBook)
DOI 10.1007/978-3-662-45704-7

Springer Heidelberg New York Dordrecht London

Printed on acid-free paper

Springer-Verlag GmbH Berlin Heidelberg is part of Springer Science+Business Media
(www.springer.com)

Foreword

The interrelation between concepts of risk, cognitive learning about uncertainty and practical risk governance is the central topic of this book written by authors from various disciplines. Plenty of authors have elaborated on each of the three topics, yet there has not been a systematic attempt to link cognitive sciences and engineering with a basic conceptualization of risk and risk governance in modern society. This integration of two distinct traditions in contemporary sciences is in itself already a major accomplishment. In addition, Jean-Marc Mercantini and Colette Faucher succeeded in linking the conceptual parts with a clear practical application for a whole set of case studies and applications.

It is a central argument of the book that through the lenses of cognitive science, at the micro level of individuals but also on the meso level of institutions and organizations, the highly disjunctive components of risk, i.e. the real experience of harm and the construction of thought experiments (what happens if) can be bridged. This insight is based on the premise that cognitions represent images of the world that have validity for understanding reality and, even more importantly, for intervening successfully in the external world (cognitive engineering). It brings to the fore a risk governance system that is, as suggested by authors, being reformulated in the face of cognitive processes of constant learning about risks during operation and practical interventions. Such a learning environment based on adaptive management is crucial for assessment as well as management of complex risks. However, Borne is sceptical about the highly generalized assumptions that cognitive processes are sufficient for understanding risks. He also includes affective and non-cognitive representations of risk sources as a major source for individual and social responses to risk issues. He develops a more sophisticated understanding of risk governance that, on one hand, reflects the knowledge acquisition process by modern institutions and, on the other hand, is compatible with the empirical research about the biases, fallacies and shortcuts of individual processing of risk information.

Examining several case studies revealed a particular relationship between open governance architecture and the ability to cope with complex risk situations. The interesting finding of the case is that the dynamics between institutional constraints and safety culture on the one side and risk management performance on the other side is more complex than previously assumed. Adaptive risk management based on cognitive engineering and thinking can destroy risk competence and create it at the same time. This ambivalence in learning structures has repercussions on the confidence in the promise of successful risk governance. It is a vital argument for choosing a precautious approach to risk management that includes the risk of inadequate risk management. So risk governance should not be envisioned as a static structure of institutional arrangements but as a permanent process, an institution under permanent construction. This interdependence of individual agency and structural arrangement reminds me of Giddens' structuration approach in which individual agency and structural constraints are closely intertwined.

The book comes to a series of very interesting and innovative conclusions. The connections between cognitive sciences and risk governance are more pronounced as one might have assumed but they are also not linear but complex. The way that institutions handle uncertainty and ambiguity prepares their thinking towards cognitive representations of the world and opens the alley for a process of adaptive management. But also the opposite seems to be true: a preoccupation for cognitive engineering might narrow the perspective for the social and cultural conditions of risk and impede effective and socially compatible risk governance. There is an analogy when one looks at the literature on resilience and robustness in risk management. Quantitative risk assessment often provides the illusion that whatever can be calculated can be managed. However, since cognitive representations are often bounded by what people select unlikely developments with high catastrophic potential are mostly ignored or underestimated.

Contributors of the book were not only able to provide a stimulating and scholarly piece of analysis; it is also an excellent reflection of the interconnections between different contemporary public discourses on risk, risk perception and cognitive sciences. Last not least the book points out to some normative implications of their analysis which should be a mandatory reading for all people in risk management and regulation.

Stuttgart, July 2014   Ortwin Renn

Contents

Chapter 1
Introduction

Jean-Marc Mercantini

Abstract The introduction chapter is a discussion about the title of the book. Why having associated the concepts of Risk and Cognition? What are the implications of this association? The responses to these questions have strong consequences on how to apprehend critical problems that emerge (or could emerge) within various activity domains. The chapter presents the two concepts of risk and cognition, and it highlights and analyses relations linking them. The complexity of the risk concept is tackled via historical, ontological and conceptual approaches. The cognition concept is defined and the filiation bonds between Cognition, Cognitive Science, Cognitive Engineering and Knowledge Engineering are presented. The set of cognition concepts defines a coherent field of interdisciplinary knowledge (scientific, methodological and technical), which provides operational tools for the analysis and the design of complex systems (or organizations) where risks are prominent.

The acceleration of the human activity evolution in a context of globalization, open markets, and scientific and technical progress, leads to in-depth changes of modern societies, and is responsible for increasingly complex situations. These situations may involve the cooperation and the collaboration of individuals from diverse cultures (e.g. social background, professional background, academic background, nationality) as well as the cooperation and the collaboration of individuals with "intelligent machines". In parallel to the acceleration of these changes, modern societies show an increased aversion to the new risks that emerge from professional, domestic and leisure activities. In this context, risk situations inherit this complexity and they may be considered to be the results of deviations from expected behaviours of systems, which involve humans, technical components, organizations and a specific environment. Their apprehension and mastery require understanding the human behaviours (individual or group basis) facing risks, and their study requires therefore mobilizing behavioural sciences, social sciences, life sciences and

J.-M. Mercantini (✉)
Laboratoire Des Sciences de L'Information et Des Systèmes, Domaine Universitaire
de Saint Jérôme, Avenue Escadrille Normandie-Niemen,
13397 Marseille Cedex 20, France
e-mail: jean-marc.mercantini@Lsis.org

© Springer-Verlag Berlin Heidelberg 2015
J.-M. Mercantini and C. Faucher (eds.), *Risk and Cognition*,
Intelligent Systems Reference Library 80, DOI 10.1007/978-3-662-45704-7_1

engineering sciences. The objective of this introduction chapter is to present the risk and cognition concepts and to highlight and analyse relations linking them. The first part of the chapter aims to identify the contours of the risk concept. The literature review highlights a concept of great complexity being the subject of a large number of research works in a variety of disciplines such as philosophy, psychology, sociology, economics, mathematics and engineering sciences. This strong mobilization of sciences is consistent to the growing anxieties of modern societies facing a future full of uncertainties and unknowns, and the feeling of not having the necessary tools for its control. Each of these disciplines brings different and complementary views of the risk concept, revealing different aspects of its complexity and its links with cognition. The second part of the chapter is dedicated to the concept of cognition as well as to sciences that study it and take it into account to design and build new systems more efficient and smarter. On one hand, considering the relationships linking the concepts of risk and cognition and, on the other hand, considering the operational tools proposed by cognitive engineering for designing and developing systems, it is possible to envisage new systems where risks may be considered as "natural phenomena", and may be taken into account in the early stages of the design process, to provide these systems with capabilities for anticipation, adaptation and resilience. The third part of the chapter presents the topics covered by this book and the way they are treated by the authors. Each chapter is described by a short abstract that gives a preview of the tackled problems and implemented methodologies to elaborate coherent solutions.

1.1 Complexity of the Risk Concept

1.1.1 Overview and Definitions

If the term risk is old and commonly used nowadays by everyone, the concept of risk comes from the probability theory, an axiomatic system derived from game theory in France in the seventeenth century [1]. The study of risk as an applied science began in the late 1960s as a response to growing public concern with new technologies [2] and recently, as a synthesis of the collective awareness in modern societies, risk and more precisely risk management has led to an international standard in 2009 [3].

1.1.1.1 The Origins of the Term

The term "risk" appeared in Occident in the twelfth century. The first text that demonstrates its use is dated of the 4th April 1248 and concerns the maritime insurance field [4]: "debet dicta navis in mari ad meum **risicum** et fortunam de omni casu".

The etymological origin of the word is not well known and linguists are not unanimous. Among many hypotheses, three origins seem to gather the largest adhesions and may be considered [4, 5]:

1. The Latin etymology, from the verb "*secare*" (to cut) in reference to hazardous marine reefs that could cut the ships' hull,
2. The Arab etymology, from the word "*rizq*" meaning "fortuitous and unexpected gift (from God)". The "rizq" can be good or bad,
3. The Roman etymology, from the verb "rixicare" (to quarrel, to fight) where running a risk means, "Pursue a dangerous and random quarrel".

With the Arab etymology, the risk is governed by god in a religious context, and with the latin etymology, the risk is under human control in an economical context.

1.1.1.2 The Origins of the Concept

The probability theory has its roots in the sixteenth century with the analysis of games of chance by Gerolamo Cardano and in the seventeenth century with Pierre de Fermat and Blaise Pascal, which are universally recognized as the founders of probability theory through the correspondence they exchanged in 1654. The first publication on a risk theory was the work of Daniel Bernoulli in: "Specimen theoriae novae de mensura sortis" (1738), where risk is defined as the mathematical expectation of a probability function of events. This definition is still widely used for quantitative risk assessment.

1.1.1.3 The Risk as an Applied Science

In the course of 1960, the environmentalist movement grows and spreads in the United States marking an increasingly deep distrust of society in respect to the development of new technologies such as nuclear. This is the first major loss of confidence that occurs between the lay public and experts. The lay public is convinced that experts minimize risk and experts blame the lay public to exaggerate [6]. This climate of distrust can be seen as a healthy reaction of society towards what can be described today as a genuine new industrial revolution taking place in the course of the sixties and the seventies. Indeed, the levels of production of the oil industry and organic chemistry are respectively multiplied by 4 and 5, while nuclear production is multiplied by 100 in France between 1973 and 1985 [7]. This increase in industrial production is accompanied almost naturally by an increase of the frequency and severity of accidents. Some of them have strongly marked the public: Flixborough (UK, 1974), Seveso (Italy, 1976), Three Mile Island (USA, 1979). The social climate of distrust, which then settles in our democratic societies, has made inevitable the systematization of risk analysis and risk management, which then have been generalized in all domains of industry. The early applications of safety analyses were mainly voluntary and based on individual companies' motivation and benefit they perceived [8]. From the late seventies, social pressures and new directives (like the EC Seveso directive) have spurred legislation, increasing the number of safety and risk analyses likely to be carried out on an authority's initiative [8]. These analyses became essential for the development and management of industrial projects.

This opposition process between the lay public and experts refers us to the work of Reinhart Koselleck [9] showing that the ability of a society to make critics, to demand accountability, to submit to a justification or judgment private or public actions, is what underlies our democracies. Three poles are essential to build critics: stakeholders dealing with systems, intellectuals and a space for public communication dominated by political representatives or spokespersons [10].

1.1.1.4 The International Standard about Risk Management

This international standard [3] describes in detail the process of risk management of any organization (any private or public company, any community, any association, any group or individual), on the basis of principles that have to be applied to make this process more effective. It applies throughout the life cycle of an organization, in all its activities and for any type of risk. The proposed generic approach provides principles and guidelines for managing any kind of risk in a systematic, transparent and reliable manner.

1.1.1.5 Definitions

Risk is a polymorphic concept that has been a focal topic of many disciplines, professional activities, and practical actions. Ortwin Renn in [11] summarizes the main disciplinary approaches to understand and analyse risks, and he highlights that "all concepts of risk have one precondition: the contingency of human actions, … and one element in common: the distinction between reality and possibility". The term risk denotes the likelihood that an undesirable state of reality (adverse effects) may occur as a result of natural events or human activities [11]. This opposition between reality and possibility is also found in Bernard Couturier [12], which suggests that "the risk is the result of an implicit comparison between what is expected, what is foreseen, what is conceptualized, and the real result that will happen". It is noticeable that these definitions are very close to that proposed by the ISO/FDIS 31000:2009 international standard about Risk Management [3] where risk is defined as the "effect of uncertainty on objectives". Where an effect is a deviation from the expected (positive and/or negative) and objectives can have different aspects (such as financial, health and safety, and environmental goals) and can apply at different levels (such as strategic, organization-wide, project, product and process).

1.1.2 Ontological Analysis of the Risk Concept

From the ontological point of view, the issues are [6]:

1. Is the risk located in the physical world with an independent existence of procedures to measure it?
2. Does the risk only exist as cognitive representation?
3. Is the risk a "hybrid" concept?

These three issues that tend to define the ontological status of the risk concept define three schools of thought who tend to oppose each other. Considering risk as a phenomenon of the physical world means to be positioned according to the positivist school of thought. Considering risk only as a cognitive representation means to be positioned according to the constructivist school of thought. Considering risk as a "hybrid" concept means to be positioned according to a current of thought where the risk is half-realistic, half-constructed.

According to the positivist thought, risk is considered as an ontological category; it has its own existence [13]. The risk is independent and external to the perceiving subject. It appears as a property of nature, technology, substance or dangerous activity. The risk is apprehended following a disembodied scientific approach, ignoring its social and cultural processes [14]. Formal and methodological tools developed according this line of thought allows to reveal a risk and therefore anticipate, prevent, compensate, etc.

According to the representational thought, the risk is not an external reality because it does not exist independently of the procedures that objectify it [15]. The risk does not exist in itself. Representations of risk are only the result of human understanding and not an accurate reflection of reality [14]. The risk is only a way to perceive and interpret particular phenomena without there necessarily being any relationship between them. Risk is a social construct, indivisible from political and cultural contingencies that base its emergence, and which alone are worthy of analyses [14].

According to the hybrid thought, theories about risk are intended to articulate both positivist and constructivist conceptions. Risk is based on reality, but it is multiple and not fixed. This reality is made visible only through the perception of risk, which is obviously socially constructed and therefore evolutive [14]. In his book "World at Risk" [16], Ulrich Beck states that "the reality of the risk is perceived through its controversial nature. Risks have no abstract existence in itself. They acquire a reality in conflicting judgments of groups and populations". The risk is seen as something real, but can be influenced by social processes [6].

In her book [6], Celine Kermisch decomposes the representational thought into three classes: the quantitative approach, the constructivist approach and the subjectivist approach.

With the quantitative representational approach, risk is conceived as the objective measure tool of potential damages, which implies the existence of something in the real world, but something that is not risk in contrast to the positivist conception. It is with this meaning that literature speaks about "objective risk" or "quantified risk". The calculation is based on the evaluation of the occurrence of events and their consequences.

With the subjectivist representational approach, risk is conceived as an individual elaboration just like individual fears. This is the sense in which literature refers when it speaks about "risk perception".

With the constructivist conception, risk is conceived as a collective elaboration and each individual conceives risks under the influence of socio-cultural context to which it belongs. The risk is the result of a social process, possibly in interaction with the world.

From this analysis, it appears that the positivist and hybrid conceptions assign to risk an existence status like an object in the real world while the representational conceptions assign to risk a cognitive representation status, whether individual or collective. The positivist and hybrid conceptions have the disadvantage to confuse risk with danger in the real world. In contrast, representational conceptions give to danger the status of object in the real world and reserve the status of representation for risk. In consequence, a greater potential wealth for the development of conceptual and methodological tools for the study and analysis of risk and danger is therefore easily predictable.

The set of works presented within the chapters of this book are resolutely in accordance with the representational conception of risk, which reinforces the idea of mobilizing cognitive science and cognitive engineering to understand the cognitive mechanisms implemented to face risks (individual or group basis) and provide adequate responses to make systems, organizations and companies more efficient to face risks. Responses can be of different types: regulatory, methodological, organizational, theoretical, etc.

1.1.3 The Various Conceptions of Risk

Although the definitions of risk are consistent at a high level of abstraction, at a more operational level, in specific application domains (insurance, health, engineering, communication, regulation, etc.), significant differences may appear. This is particularly the case concerning the characteristics of the systems subject to the risk analysis, the implemented methodological approaches, and the valuation models. Many authors [2, 6, 11, 17, 18] have worked to classify these different perspectives whose summary is presented in this paragraph.

1.1.3.1 The Quantitative Representational Conception of Risk

According to this approach, risk is conceived as the objective measure of potential damages and the measuring tool depends on the perspective of the risk analysis. In reference to the Renn classification [11, 18], the following three perspectives can be subsumed under the quantitative representational conception: (i) the actuarial perspective, (ii) the Causal Modelling perspective, and (iii) the Probabilistic Risk Assessment (PRA) perspective.

The actuarial perspective consists in applying mathematics and statistics in the domain of insurance where the risk-measuring tool is the statistical expectation value of financial losses.

The Causal Modelling perspective is a modelling process, which consists in determining causal relationships between dangerous phenomena and physical potential damages to humans or others living organisms. The risk-measuring tools are the results of this modelling process and they are specifics to the dangerous phenomena and the physical potential damages. Such causal models are developed for measuring chronic and accidental risks and they are based on toxicological or epidemiological studies.

The PRA perspective is a modelling process, which consists in modelling the dependability (defined in terms of reliability, availability, maintainability and safety) of complex technological systems. The objectives of these models are to point out safety-relevant failures induced by technical faults and/or human errors and to determine their consequences. Each combination of negative events (faults or errors) which leads to safety-relevant failure combined with possible consequences, constitutes an accident scenario. The probabilistic quantification of the different accident scenarii gives a measure of the risk. The modelling process is made of the implementation of several predictive analysis methods such as Failure Mode Effect and Criticality Analysis (FMECA), Fault Tree Analysis (FTA), Event Tree Analysis (ETA), Risk Matrix combined with the use of Geographic Information Systems (GIS) for spatial analysis (location of hazards and stakes); and the use of Human Error Analysis, Human Reliability Assessment (HRA) and empirically driven human-machine interface simulation to take in account human behaviours within complex technological systems.

1.1.3.2 The Subjectivist and Constructivist Conceptions of Risk

With the subjectivist and constructivist conception, individuals, groups of individuals and society are major components for risk assessment. In reference to the Renn classification [11, 18], the following four perspectives can be subsumed under the subjectivist and constructivist conceptions: (i) the economic perspective, (ii) the psychological perspective, (iii) the sociological perspective and (iv) the cultural perspective. In contrast with the objective quantitative conception of risk, undesired effects are not limited to financial losses or physical harms but they are extended to all effects leading to consequences for something that people value [11].

The economic perspective

The economic perspective considers the decision-maker's point of view where risks have to be quantified, to be easily compared and prioritized. However, the measure

of risk by means of the expectation value is problematic for at least two fundamental reasons [2]: (i) probability-weighing is normatively controversial and (ii) it
assesses risks only according to their probability and the severity of their consequences. For these reasons, the economic approach is using the expected utility
instead of the expected value, as a risk measuring tool. The introduction of expected
utility functions give the opportunity to the decision maker's to consider a degree of
satisfaction (or dissatisfaction) associated to possible options (a probability distribution function is associated to the set of possible options). Expected utility
functions are specifics to individual (or decision maker). They constitute the formal
representation of subjectivity.

However, the paradoxes raised by the experiments of Allais and Ellsbergs [19, 20]
about the expected utility model has been the source of new fields of investigation
especially among researchers in cognitive psychology to better understand the
behaviour of the decision maker under risk or uncertainty [6].

The psychological perspective

Herbert A. Simon shows that the expected utility model is inadequate to describe
the decision-making process because the set of options is not given but developed
by the decision maker himself [6]. It therefore becomes essential to analyse and
understand the cognitive development process. Herber A. Simon proposes the
theory of the "bounded rationality". The main idea is that in decision-making,
rationality of individuals is limited by the information they have, the cognitive
limitation of their minds, and the finite amount of time they have to make a
decision. The consequence is that the decision-maker is unable to seek the optimal
solution but a satisfactory one; he becomes a "satisficer". Herbert A. Simon
introduced the "satisficing" concept [21], which is the combination of the terms
"satisfy" and "suffice", to describe the decision-making strategy to meet an
acceptability threshold.

In the course of the 1970s, Daniel Kahnemann and Amos Tversky provide
empirical proofs that choices within risky or uncertain situations systematically
violate the axioms of expected utility theory. They argue that utility theory, as it
was commonly interpreted and applied, was not an adequate descriptive model [22].
They propose that "when faced with the difficult task of judging probability or
frequency, people employ a limited number of heuristics which reduce these
judgements to simpler ones". Unfortunately these heuristics can lead to serious and
systematic errors called "cognitive biases" [23, 24]. Ortwin Renn in [11] presents
four cognitive biases of risk perception: the availability bias, the anchoring effect
bias, the representation bias and the avoidance of cognitive dissonance bias.

In 1979, Kahneman et Tversky [22] propose the "prospect theory", a more
accurate and realistic description of the decision making process than the expected
utility theory. It describes the asymmetric evaluation of the loss and gain prospects.
People are risk-averse if they face potential losses, and risk-prone if they expect
even small gains [11].

Critiques and questioning of the expected utility theory has fostered the emergence and development of researches about risk perception. Although risks perception differ considerably among social and cultural groups, it appears to be a common characteristic that most people form their beliefs by referring to the nature of the risk, the cause of the risk, the associated benefits, and the circumstances of risk-tacking [11].

Within a social climate of distrust between lay public and experts (during 1970s), and the deficiencies of the probabilistic models, Paul Slovic, Baruch Fischoff and Lichtenstein propose, on the basis of cognitive biases, the psychometric paradigm whose goal is to characterize risk perception among lay public. The psychometric paradigm has highlighted two major results: (i) perceived risks and perceived benefits (related to an activity or a technology) are evolving in opposite directions (higher are estimated benefits, lower are perceived risks) and, (ii) perceived risks and their acceptation are based on a set of qualitative attributes (such as "dreadful", "threat for future generations", "familiarity", "controllable", "personal exposure", etc.) [25–27]. The works of Fischoff and Slovic show that people develops a very acute perception of risks based on the consideration of these qualitative attributes. However, two important assumptions set limits of the psychometric paradigm: (i) risk perception is individual and independent of any social or cultural representation, (ii) all individuals react similarly when facing the same risk.

The psychological perspective on risk includes all undesirable or desirable effects that people associate with a specific cause (no restriction) and probabilities are substituted by the strength of belief that people have about the likelihood [11]. According to Fischhoff [28], risks perception studies contribute to improving risk policies in such way [11]: (i) they reveal public concerns and values; (ii) they serve as indicators for public preferences; (iii) they document desired lifestyles; (iv) they help design risk communication strategies and, (v) they represent personal experiences in ways that may not be available to the scientific assessment of risk.

The cultural perspective

The cultural theory of risk (early 1980s) is based on the cultural theory initiated by Mary Douglas from her early works about the perception of the dirt within primitive societies [29]. The cultural theory of risk wants to propose a response to the question: "why people do not perceive the same risks in the same way?" [6, 30].

The term "culture" means all of beliefs, values, ways of perceiving the world and react to it [6]. Individuals make meaning of their actions according to the requirements of the social context in which they operate and the perception of risks is linked to the "cultural prototype" (typical combinations of values, world views, and conviction) in which they emerged. Cultural prototypes identify cultural groups in society with specific positions on risk topics as well as corresponding attitudes and coping strategies [11]. Cultural prototypes are characterized along two independent dimensions: the group cohesiveness dimension (called "group") and the "grid" dimension. The group dimension measures how much of people's lives is controlled by the group they live in (social incorporation), and the grid dimension measure the amount of control their members accept (role differentiation within a system of hierarchy) [30].

Considering these two dimensions, Mary Douglas in Cultural bias [32] proposes four types of social forms (or cultural prototypes) characterized by the combination of the two possible values (weak or strong) taken by each dimension: (i) the positional form (or hierarchical, strong group and strong grid), (ii) the enclave form (or egalitarian, strong group and weak grid), (iii) the individualist form (or entrepreneur, weak group and weak grid) and, (iv) the isolate form (weak group and strong grid). Each of these cultural forms is characterized by a cultural bias, which in turn influences the perception (or the construction) of risks. There is no true or false perception, but cultural biases that condition perceptions. Within the positional social form, cultural bias supports tradition and order [31] and risks are acceptable as long as institutions have the routine to control them [11]. Within the enclave social form, cultural bias supports equalities, solidarity, rejects the outside world [31] and risks should be avoided unless they are inevitable to protect the public good [11]. Within the individualist social form, cultural bias supports competition and private benefits [31] and risks offer opportunities and should be accepted in exchange for benefits [11]. Within the isolate social form, cultural bias supports isolation, apathy, dependencies [31], and risks are out of control and safety is a matter of luck [11].

The sociological perspective

The sociological perspective is complex in itself for at least two reasons: (i) there are as many perspectives within sociology as there are sociologists [18] and (ii) the political dimension of the issues raised by risk in our modern societies [6]. Issues that were initially confined within the domains of experts or scientists have moved in the public domain and hence, have led to new relationships between individuals and new management of society [6]. To illustrate the sociological perspective, the reflexive modernization theory and the social amplification theory will be presented.

Principles of the reflexive modernization theory are based on the works of Ulrick Beck [33] and Anthony Giddens [34, 35]. The risk is a concept revealing the essential features of the contemporary society [6]: individualization of lifestyles and social careers, pluralisation of knowledge camps and values, lack of overarching objectives and goals, experience of negative side effects. Risk is the central focus of debate and controversy [11] and according to Beck, risk is the ultimate link that connects individual to society where the allocation of wealth is substituted by risk allocation. In his book, *Risk society* [33], Ulrick Beck describes modern society as a safe state paradoxically threatened from interior by the risk. Jacques Theys [7] had also noted this paradox of modern societies where increasing security only reinforces the public aversion to technological risks.

The aim of the theory of the social amplification of risk is to understand "why some relatively minor risks or risk events, as assessed by technical experts, often elicit strong public concerns and result in substantial impacts upon society and economy" [36]. It consists in the study of the process leading to the transformation of the risk perception. The theory was developed in the late 1980s and it is integrating the technical analysis of risk and the cultural, social and individual response structures that shape the public experience of risk [36]. Risk is defined as the result of two components: an objective threat of harm to people and a product of culture and social experience. The main hypothesis of the theory is that risk events interact with psychological, social, institutional and cultural processes in ways that can heighten or attenuate public perceptions of risk and shape risk behaviour [37]. These perceptions and behaviours generate social and economic indirect consequences that can be significant and in turn will result in new phenomena of amplification or attenuation. The process is iterative. This theory puts communication at the heart of the processes of risk perception, amplification and attenuation. The principles of this theory come from the communications theory. The amplification denotes the process of intensifying or attenuating signals during the transmission of information from an information source to intermediate transmitters, and finally to a receiver [38]. In social communication the process is quite more complex than in electronic communication because sources, transmitters, receivers and information are not independent entities [36].

1.2 From Cognition to Engineering

The purpose of this paragraph is to define the concepts of Cognition, Cognitive Science, Cognitive Engineering and Knowledge Engineering, as well as to clarify the filiation bonds between them. These concepts define a coherent field of interdisciplinary knowledge (scientific, methodological and technical) where Cognitive Engineering and Knowledge Engineering provide operational tools for the design of complex systems in which humans and intelligent machines are working together (in a cooperative and collaborative way) to accomplish complex missions.

1.2.1 Cognition

The term Cognition is coming from the latin verb "Cognoscere" meaning "to become acquainted with", "to come to know". Its first known use is during the XIVth century in French and during the XVth century in English.

From Merriam Webster encyclopaedia, Cognition is the act or the process of knowing. It includes every mental process that may be described as an experience of knowing (including perceiving, recognizing, conceiving, and reasoning), as distinguished from an experience of feeling or willing. From the cognitive science dictionary [39], cognition is a function allowing the knowledge realisation and examining the different activities relating to knowledge. Jean-Gabriel Ganascia [40] defines cognition as the study of knowledge; it evokes the meeting of disciplines that deal with knowledge in any areas of concern, its sources, its supports and its vehicles. Cognition may be also defined as the ability to integrate multimodal information for generating representations, building associations and elaborating generalizations. The ability to manipulate this knowledge allows the individual to develop a behaviour that depends not only on the environment or the immediate situation.

Originally, the sciences of cognition are based on the study of natural cognition for then evolving toward the study of artificial cognition mobilizing computers to reproduce the mental representations and the functions that allow their treatment. Cognition became an object of scientific study during the twentieth century. Its development is strongly linked to the development of computers used as tools to simulate the cognitive process models, but also used as a metaphor of the brain function where information is received, formatted, processed and stored in memory. This memory is then mobilized to elaborate reasoning and action plans.

Various disciplines, such as psychology, philosophy, linguistics, computer sciences and sociology are studying cognition but the concept is covering different semantic fields. In psychology and cognitive science, cognition usually refers to an information processing or a mental representation of individuals. In social psychology (or social cognition), the cognition concept is used to explain attitudes, attribution, and group dynamics [41].

1.2.2 Cognitive Science

Cognitive Science (CS) can be defined as the interdisciplinary scientific study of intelligence and its computational processes in humans (and animals), in computers, and in the abstract [42]. CS is at the intersection of information science, life sciences and human sciences and, it examines what cognition is, what it does and how it works.

Cognitive science emerged in 1956 from the early development of the cybernetics, the theory of computation and the digital computer. The founding ideas come from Warren McCulloch and Norbert Wiener. McCulloch wanted to explain mental phenomena by the neuroanatomical organization of the brain and Wiener worked onto similarities between living systems and artificial automata. The fields that contributed to the birth of cognitive science are philosophy, linguistics, anthropology, neuroscience, computer science, and psychology.

Since the emergence of cognitive science, three founding revolutions may be distinguished [43]:

1. The period in which the central role is given to cognitive psychology. From there, emerged the concept of mental representation.
2. The period in which it was noticed that mental representation and language are inseparable. This result was used to simulate the mental representation by computer programs.

3. The period in which the physical substrate of mental representation in the brain has been discovered. The study of this physical substrate (by means of neuro-imaging) has made possible the correlation of brain activity with cognitive activity and behaviours.

With cognitive science, the understanding of the outside world changes its viewpoint. It is not external objects that attract attention, but the tool with which they are observed. Cognitive science is concerned with the study of the processes of perception, reasoning, pattern recognition, concept formation, understanding, interpretation, problem solving, control, planning and action. Cognitive engineering and knowledge engineering will propose formal methods, guidelines and norms to design systems in which cognition has a central position.

1.2.3 Cognitive Engineering

From the Oxford Handbook of Cognitive Engineering [44], Cognitive Engineering is an interdisciplinary approach to the analysis, modelling, and design of engineered systems or workplaces, especially those in which humans and automation jointly operate to achieve system goals. From Vicente [45], Cognitive engineering is a multidisciplinary endeavour concerned with the analysis, design, and evaluation of complex systems of people and technology.

Both definitions are very close and they characterize an area of activity (scientific and technical) that is concerned by integrated human-technology systems. It combines knowledge and experience from Cognitive Science, Human Factors, Human-Computer Interaction Design and Systems Engineering [46]. Cognitive Engineering emerged in the early 1980s in response to transformation in the workplace by two major sources [46]: (i) computer systems were escaping from the confines of machine rooms and thus design principles were needed to ensure than ordinary people would be able to use them and, (ii) Safety Critical Systems were becoming more complex and increasingly computer controlled; design principles were needed to ensure that teams of skilled technicians could operate them safely and efficiently. Otherwise, this emergence is also linked to the maturation of cognitive science into a discipline whose theories, models and methods are capable of guiding application.

1.2.4 Knowledge Engineering

Knowledge Engineering is the scientific field that studies the process of building Knowledge Base Systems (KBSs) with the objective of developing tools, such as concepts, principles, techniques, methods, languages and software, to help design high quality KBSs [47]. KBSs are exploited to achieve or help achieve knowledge-intensive human tasks, generally difficult to formalize and for which modelling and acquisition of knowledge are generally required [48]. KBSs are used to capitalize, produce and share knowledge within a community of users or experts [49–51], knowledge being considered an important asset [52–55].

One goal of KBSs is the use of knowledge for solving or helping solve complex problems in a chosen domain. However, the first generation of KBSs failed in doing this task, being not sufficiently robust [56]. They were often modelled at a too low level of abstraction, most often at the symbol level, using an implementation language. Indeed, knowledge was often said to be extracted from the expert discourse (usually expressed at the linguistic level) and directly translated in an implementation language. Among the solutions proposed with the second generation of KBSs [56] were conceptual modelling, ontology and problem-solving methods (PSMs) [47, 56].

Conceptual modelling was proposed to explicitly elaborate models from the expert discourse, without implementation details. The underlying idea of a conceptual step was to model at a more abstract level (the knowledge-level) [57] the knowledge contained in the expert discourse (knowledge sources being interviews, corpus, etc.) in the process of solving complex problems of the domain. Conceptual modelling was seen as an understandable intermediary step before formalization and implementation. The conceptual model lies between human expertise and the implemented program, and this model determines the construction of the formal knowledge base [58]. This is in this context that, since the middle of the 1980 years, several methods have been elaborated to help modelling KBS at the conceptual level, for instance Knowledge Oriented Design (KOD) [59], KADS [60], CommonKADS [53, 61], Protégé [62], MIKE [63] and MACAO [64].

Conceptual models can be elaborated with descendant methods (also called top-down or model-directed, for instance KADS and CommonKADS) by specializing models, ascendant methods (also called bottom-up or data-directed, for instance MACAO and KOD) by abstracting data to model, or middle-out methods that abstract and specialize the most important concepts. The KADS method proposed the concept of *expertise model* [60] to structure the knowledge needed to solve problems (by an expert or KBS), concept later recalled *knowledge model* in CommonKADS [53, 61]. One purpose of this kind of model is to distinguish the types of knowledge needed to solve problems: domain, inference and task. The expertise model is an example of a knowledge level model [65].

The concept of *PSM* can be defined as a knowledge-level description of a problem-solving process, making abstract of implementation details [47]. In other words, a PSM refers to the reasoning made by the expert when solving a given problem. An anticipated KBS might solve a problem with one or several selected PSMs. Problem solving knowledge is distinguished from domain knowledge [65]. A task model in CommonKADS [66] is an example of problem solving knowledge.

1.3 Summary of the Book

The following five topics are covered by the book:

- Influence of the culture in risk management,
- Influence of the risk communication in risk management,
- User-centred design to improve risk situation management,
- Designing new tools to assist risk situation management,
- Risk prevention in industrial activities.

The topic *influence of the culture in risk management* is illustrated by the Chap. 2 where the author shows that the same phenomenon can be perceived and experienced differently by different nations according to their cultural characteristics expressed in terms of tolerance to uncertainty and risk. For his demonstration, the author analyses the policy divide between north America countries, south America countries and the countries of the European Union, concerning the problem of genetically modified food. The analysis asks whether national differences in political culture, as expressed through different levels of tolerance for uncertainty and risk affect the formulation of protective regulatory policy. The study hypothesizes that varying levels of uncertainty tolerance coupled with prevailing risk perceptions either encourage the implementation of new protective policies or lead to the adjustment of existing regulations.

Two chapters illustrate the topic *influence of the risk communication in risk management*. In the Chap. 3 the authors present the case of a country (Serbia) where the risk communication system shows many deficiencies from operational and academic point of view. One of the results of the analysed situation is that message processing depends on the cognitive characteristics and cognitive limitations of the message recipients, as well as message attributes. It is of prime importance that policy makers keep these two considerations in mind. The most critical result observed in this case study is when citizens are not any more in trust with their authorities. In the Chap. 4 the author presents a scientific approach to understand the impact of some forms of communication onto the social cognitive representation of risks. To illustrate her demonstration, the author treats the problem of the communication within foodborne crisis situations. The study concentrates on verbal and nonverbal metaphors and their role in risk cognition. The author studies the metaphorical dimension of risk cognition as well as its dynamics connected with the necessity of constant response to the changing internal and external conditions. The aim of this research is to show whether metaphors strengthen or weaken risk cognition and how to determine the risk communication of foodborne diseases.

Two chapters are dedicated to the topic *User-centred design to improve risk situation management*. In the Chap. 5 the authors tackle the problem of risks induced by the use of artefacts resulting from a poor design. Poor design is believed to increase risks. The adoption of User-Centred Design (UCD) can be seen as preventive and protective measures to reduce risks of the artefact users. UCD becomes a risk management tool. The adoption of UCD obliges designers to take in account subjective risk instead of being focused on the objective risk. To design safe products it is necessary to acknowledge that it is the subjective perceptions of an individual that will influence their future actions, rather than any real consideration of the objective risk. Building an artefact for the end-user instead of building an artefact with the hypothesis that end-users will adopt the designer vision. In the Chap. 6, authors present a method to investigate the human behaviour and its relation with accidents and human errors. The method is supported by an experimental protocol for the observation of the user interaction based on a multi-disciplinary model, which involves the understanding of the individual's within work situations. The human error is studied from a cognitive psychology perspective and the work situations are implementing interactive systems (artefacts) during critical situations. The final objective of this chapter is to demonstrate how raising the quality level of the human interaction in critical situations can contribute to risk management.

The topic *designing new tools to assist managers in at-risk situation* is about the use of cognitive science to solve complex problems involved in managing at-risk situations. Two chapters are dedicated to this topic: the first treats the fault diagnosis problem and the second treats the planning problem of emergency actions. In the Chap. 7, the authors tackle the cognitive process of the diagnosis by means of a formal multi-modelling method and a diagnosis algorithm. The multi-modelling method (called Timed Observations Method for Diagnosis: TOM4D) is based on the elaboration of four models: a Structural Model, a Functional Model, a Behavioural Model and a Perception Model. The resulting process allows the automatic fault detection, identification and diagnosis and it is applied to hydraulic dam safety. In the Chap. 8 the author shows that cognitive approaches can offer very powerful engineering environments to tackle issues raised by risk management. The investigated issue is the planning of actions to fight accidental marine pollutions. The response proposed is a software tool to support stakeholders to plan fight actions during emergency situations or crisis management with the objective to minimize pollution impacts. From a methodological perspective, the chapter shows the importance to develop ontologies (i) for structuring a domain as perceived by its actors and (ii) for building computer tools aimed to support problem solving in that domain. Such tools are imprinted with the knowledge shared by the actors of the domain, what make them more effective within critical situations.

The topic *risk prevention in industrial activities* is illustrated by the Chap. 9 where authors analyze the working activity of professionals in charge of safety in industrial companies (preventionists). The purpose of this activity consists in adapting the regulations relative to industrial activities, from texts of law of general order towards their implementation in a specific context. This process has been analyzed by the authors according to the instrumental approach of Rabardel, which present interesting properties. In particular, the systems of instruments are structured according to the experience and skills of the workers, and they are characterized by complementarities and functional redundancies, following the example of a security system. This approach takes into account the elements of the context, the constraints and the resources of the activity. A case study is presented where the regulations function as a "pivot instrument" of this system. From the regulations, the preventionist establishes the diagnosis of the company safety level and develops the corresponding preventive and formative actions.

1.4 Conclusion

The literature review about the risk concept has revealed a universe of knowledge that has evolved in complexity and accuracy depending on human thoughts and the increasing complexity of modern societies. Psycho-cognitive, cultural and psycho-social approaches have had to be considered to better understand the perception of this concept by individuals, social groups or societies. The concept of risk cannot be reduced to a simple mathematical formula.

The literature review about cognitive science has led to discover a scientific field that has expanded considerably since its inception in 1956. It covers the spectrum from the study of psychological phenomena up to the knowledge engineering, which offers scientific approaches and methodological tools to design artefacts that have to be associated with human beings. This association forms couples increasingly indivisible where: artefacts inherit human intelligence and human beings change their worldviews, their behaviours and their social organizations depending on these artificial beings.

The combination (or the connection) of the risk concept with cognition and cognitive science leads almost "naturally" to the idea of building new intelligent systems where human beings and artefacts can work together in a coherent organization to face risks. It implies new approaches and new tools to model, to analyse, to control, to predict, to prevent and to protect. But risks are also cognitive constructions and it is also necessary to communicate, to sensitize, to concert, to collaborate, to train, to organize and to regulate. The joint consideration of risk and cognition leads to address risk issues with a more comprehensive and coherent vision, which may lead to the design of new tools marked of consistency.

References

1. Douglas, M. (1987). Les etudes de perception du risque: un état de l'art. *La société vulnerable* (pp. 55–60). Paris: Presses de l'École Normale Supérieure.
2. Hansson, S. O. (2005). The Epistemology of technological risk. *Techné, 9*(2), 68–80.
3. ISO/FDIS 31000:2009. Risk management—Principles and guidelines.
4. Villain-Gandossi, C. (1990). Origines du concept de risque en Occident. Les risques maritimes ou de fortune de mer et leur compensation: les débuts de l'assurance maritime. In L. Faugères, P. Vasarhelyi, & C. Villain-Gandossi (Eds.), *Le risque et la crise* (pp. 71–84). European Coordination Centre for Research and Documentation in Social Sciences, ISBN 3-900815-10-0.
5. de Epalza, M. (1990). Origine du concept de risqué de l'Islam à l'Occident. In L. Faugères, P. Vasarhelyi, & C. Villain-Gandossi (Eds.), *Le risque et la crise* (pp. 63–70). European Coordination Centre for Research and Documentation in Social Sciences, ISBN 3-900815-10-0.

6. Kermisch, C. (2011). *Le concept de risque: de l'épistémologie à l'étique*. Paris: Editions Tec and Doc, Lavoisier.
7. Theys, J. (1987). La société vulnerable. *La société vulnerable* (pp. 3–36). Paris: Presses de l'École Normale Supérieure.
8. Suakas, J., & Rouhiainen, V. (1993). *Quality management of safety and risk analysis*. Amsterdam: Elsevier.
9. Koselleck, R. (1979). *Le règne de la critique*. Paris: Éditions de Minuit.
10. Chateauraynaud, F. (2003). Redoubler de vigilance: les contraintes cognitives et les enjeux politiques des nouveaux modèles de gestion des risques. In *Réflexions autour du risque: Définitions, Prévention et Évolution* (pp. 51–59). Journées d'études ENAP, CIRU, CIRAP.
11. Renn, O. (2008). Concepts of risk: An interdisciplinary review (Part1: Disciplinary risk concepts). *GAIA, 17*(1), 50–66.
12. Couturier, B. (2003). Le risque: de l'approche philosophique à l'approche anthropologique. In *Réflexions autour du risque, Définitions, Prévention et Évolution* (pp. 41–50). Journées d'études ENAP, CIRU, CIRAP.
13. Rescher, N. (1983). *Risk: A philosophical introduction to the theory of risk evaluation and management*. Washington D.C.: University Press of America.
14. Pieret, J. (2012). Épistémologie du risque: la troisième voie d'Ulrick Beck et son influence sur la doctrine environnementaliste. Lex Electronica, (Vol. 17.1), Summer 2012.
15. Caeymaex, F. (2007). Risquer, gérer, sécuriser: Techniques politiques de la modernité? In C. Kermisch & G. Hottois (Eds.), *Techniques et philosophies des risques* (pp. 111–122). Paris: Vrin.
16. Beck, U. (2009). *World at risk*. Cambridge: Polity Press.
17. Kermisch, C. (2012). Vers une definition multidimensionnelle du risque. http://vertigo.revues.org/12214.
18. Renn, O. (1992). Concepts of risk: A classification. In S. Krimsky & D. Golding (Eds.), *Social theories of risk* (pp. 53–79). Westport: Praeger.
19. Allais, M. (1953). Le comportement de l'homme rationnel devant le risque, critique des postulats et axiomes de l'école américaine. *Econometrica, 21*, 503–546.
20. Ellsberg, D. (1961). Risk, ambiguity and the savage axioms. *Quaterly Journal of Economics, 75*, 643–669.
21. Simon, H. A. (1956). Rational choice and the structure of the environment. *Psychological Review, 63*(2), 129–138. doi:10.1037/h0042769.
22. Kahneman, D., & Tversky, A. (1979). Prospect theory: An analysis of décision under risk. *Econometrica, 47*(2), 263–292.
23. Tversky, A., & Kahneman, D. (1973). Availability: A heuristic for judging frequency and probability. *Cognitive Psychology, 5*, 207–232.
24. Tversky, A., & Kahneman, D. (1974). Judgement under uncertainty: Heuristics and biases. *Science, 185*, 1124–1131.
25. Fischoff, B., Slovic, P., Lichtenstein, S., Read, S., & Combs, B. (1978). How safe is safe enough? A psychometric study of attitudes toward technological risks and benefits. *Policy Sciences, 9*, 127–152.
26. Slovic, P., Fischoff, B., & Lichtenstein, S. (1980). Facts and fears: Understanding perceived risk. In R. Schwing & W. Albers (Eds.), *Societal risk assessment* (pp. 181–214). New York: Plenum Press.
27. Slovic, P. (1992). Perception of risk: Reflections on the psychometric paradigm. In S. Krimsky, & D. Golding (Eds.), *Social theories of risk* (pp. 153–178). Westport: Praeger.
28. Fischoff, B., Slovic, P., & Lichtenstein, S. (1985). Weighing the risks. In R. W. Kates, C. Hohenemser, & X. J. Kasperson (Eds.), *Perilous progress: Managing the hazards of technology* (pp. 265–283). Boulder: Westview.
29. Douglas, M. (1966). *Purity and danger: An analysis of concepts of pollution and taboo*. London: Routledge & Kegan Paul.

30. Kermisch, C., & Labeau, P. E. (2008). Cultural theory and risk perception: A critical analysis. *Proceedings du 16ième Congrès de Maîtrise des Risques et de Sûreté de Fonctionnement*. Avignon, octobre 6–10, 2008. ISBN: 2-35147-028-1.
31. Douglas, M. A history of grid and group cultural theory. Retrieved July 27, 2013, from http:// projects.chass.utoronto.ca/semiotics/cyber/douglas1.pdf.
32. Douglas, M. (1979). *Cultural bias* (Occasional paper No. 35). London: Royal Anthropological Institute.
33. Beck, U. (1992). *Risk society: Toward a new modernity*. London: Sage.
34. Giddens, A. (1990). *The consequences of modernity*. Stanford: Stanford University Press.
35. Giddens, A. (2000). *Runaway world*. London: Routledge.
36. Kasperson, R., Renn, O., Slovic, P., Brown, H., Emel, J., Goble, R., et al. (1988). The social amplification of risk: A conceptual framework. *Risk Analysis, 8*(2), 177–187.
37. Renn, O., Burns, W. J., Kasperson, J. X., Kasperson, R. E., & Slovic, P. (1992). The Social amplification of risk: Theoretical foundations and empirical applications. *Journal of Social Issues, 48*(4), 137–160.
38. DeFleur, M.L. (1966). *Theory of mass communication*. New York: David McKay.
39. Tiberghien, G. (2002). *Dictionnaire des sciences cognitives*. Paris: Armand Colin.
40. Ganascia, J. G. (1999). *Sécurité et cognition*. Paris: Hermes.
41. Sternberg, R.J., & Sternberg, K. (2009). *Cognitive psychology* (6th Ed.). Belmont: Wadsworth, Cengage Learning.
42. Simon, H. A., & Kaplan, C. A. (1989). Foundations of cognitive science. In M. I. Posner (Ed.), *Foundations of cognitive science* (pp. 1–47). Cambridge: A Bradford book, MIT Press.
43. Bayatani, M. (2007). *De la cybernétique aux sciences de la cognition*. PhD Thesis, Lyon III, Février 2007.
44. Lee, J. D., & Kirlik, A. (2013). Cognitive engineering: History and foundations. In J. D. Lee & A. Kirlik (Eds.), *The Oxford handbook of cognitive engineering*. Oxford, New York: Oxford University Press.
45. Vicente, K. J. (1999). *Cognitive work analysis: Toward safe, productive, and healthy computer-based work*. Mahwah: Lawrence Erlbaum Associates, Inc.
46. Gersh, J. R., McKneely, J. A., & Remington, R. W. (2005). Cognitive engineering: Understanding human interaction with complex systems. *Johns Hopkins APL Technical Digest, 26*(4), 377–382.
47. Studer, R., Benjamins, V. R., & Fensel, D. (1998). Knowledge engineering: Principles and methods. *Data & Knowledge Engineering, 25*(1–2), 161–197.
48. Charlet, J., Zacklad, M., Kassel, G., & Bourigault, D. (2000). Ingénierie des connaissances: recherches et perspectives. In J. Charlet, M. Zacklad, G. Kassel, & D. Bourigault (Eds.), *Ingénierie des connaissances. Évolutions récentes et nouveaux defies* (pp. 1–22). Eyrolles, 2000.
49. Neches, R. (1991). Acquisition of knowledge for sharing and reuse. In B.R. Gaines (Ed.), *Proceedings of the Annual Workshop on Knowledge Acquisition*. Banff, Canada, October 6– 11, 1991.
50. Neches, R., Fikes, R., Finin, T., Gruber, T., Patil, R., Senator, T., et al. (1991). Enabling technology for knowledge sharing. *AI Magazine, 12*(3), 36–56.
51. Caussanel, J., & Chouraqui, E. (2000). Contribution à l'étude des Systèmes de Capitalisation des Connaissances : SMOKC, un système dédié aux PME-PMI. Université d'Aix-Marseille 3, Mémoire de thèse, Aix-en-Provence, 2000.
52. Nonaka, I., & Takeuchi, H. (1995). *The knowledge-creating company*. New York: Oxford University Press.
53. Schreiber, G., Akkermans, H., Anjewierden, A., de Hoog, R., Shadbolt, N., Van de Velde, W., et al. (2000). *Knowledge engineering and management*. The MIT Press, Cambridge: A Bradford Book.
54. Ermine, J.L. (2004). Introduction au knowledge management. In I. Boughzala, & Ermine J.L. (Eds.), *Management des connaissances en entreprise* (pp. 55–77). Paris: Lavoisier, Publications Hermès Science.

55. Nonaka, I., Troyama, R., & Konno, N. (2005). SECI, ba and leadership: A unified model of dynamic knowledge creation. In S. Little & T. Ray (Eds.), *Managing knowledge. An essential reader* (2nd ed., pp. 23–49). Londres: The Open University, SAGE Publications.

56. David, J. M., Krivine, J. P., & Simmons, R. (Eds.), *Second generation expert systems* (pp. 232–272). Berlin: Springer.

57. Newell, A. (1982). The knowledge level. *Artificial Intelligence, 18*(1), 87–127.

58. Luger, G. F. (2005). *Artificial intelligence: Structures and strategies for complex problem solving* (5th ed.). Harlow: Addison-Wesley.

59. Vogel, C. (1988). *Génie cognitif*. Paris: Masson (Sciences cognitives).

60. Schreiber, G., Wielinga, B., & Breuker, J. (Eds.). (1993). *KADS: A principled approach to knowledge-based system development, knowledge-based systems* (Vol. II). Boston: Academic Press.

61. Schreiber, G., Wielinga, B., & de Hoog, R. (1994). CommonKADS: A comprehensive methodology for KBS development. *IEEE Expert, 9*(6), December, 28–37.

62. Puerta, A. R., Egar, J. W., Tu, S. W., & Musen, M. A. (1992). A multiple-method knowledge acquisition shell for the automatic generation of knowledge acquisition tools. *Knowledge Acquisition, 4*, 171–196.

63. Angele, J., Fensel, D., & Studer, R. (1998). Developing knowledge-based systems with MIKE. *Journal of Automated Software Engineering, 5*(4), 389–418.

64. Aussenac, N. (1989). *Conception d'une méthodologie et d'un outil d'acquisition de connaissances expertes*, Thèse de l'université Paul Sabatier de Toulouse, October 1989.

65. Uschold, M. (1998). Knowledge level modelling: Concepts and terminology. *The Knowledge Engineering Review, 13*(1), 5–29.

66. Breuker, J., & van de Velde, W. (1994). CommonKADS library for expertise modelling. In J. Breuker (Ed.), *Col. frontiers in artificial intelligence and applications* (Vol. 21). Amsterdam: IOS Press.

Chapter 2
The Role of Risk Perception and Political Culture: A Comparative Study of Regulating Genetically Modified Food

Tony E. Wohlers

Abstract Policymakers in industrialized countries have responded differently to the perceived opportunities and threats regarding the genetic modification of agricultural food production. In particular, a biotechnology policy divide has emerged since the 1990s between North America and some countries in South America on the one hand and many countries in the European Union. This study asks whether national differences in political culture, as expressed through different levels of tolerance for uncertainty and risk affect the formulation of protective regulatory policy in the area of genetically modified food. To answer this question, the analysis applies elements of the cultural model developed by Hofstede and uses a modified version of the Margolis Risk Matrix to assess risk tolerance in regards to the regulation of genetically modified food in the United States, Canada, Brazil, and the European Union.

2.1 Introduction

The discovery of the molecular structure of deoxyribonucleic acid or DNA by James Watson and Francis Crick opened the door for the "direct, intentional alteration of the genetic materials of organisms [by] moving genes from one organism to another" [1]. The subsequent advances in and diversification of genetic modifications of agricultural food production through the technique of genetic engineering have paved the way for the expansion of biotechnology in agriculture across the globe. While industrialized countries like the United States and Canada dominate, developing nations like Argentina, India, and especially Brazil have also become major global players in agricultural biotechnology. This global expansion of genetic applications in agricultural has also sparked debate over the benefits and risks associated with them [2, 3]. Some argue that the predictability associated with

T.E. Wohlers (✉)
Department of History and Government, Cameron University, Lawton, OK 73505, USA
e-mail: awohlers@cameron.edu

© Springer-Verlag Berlin Heidelberg 2015   21
J.-M. Mercantini and C. Faucher (eds.), *Risk and Cognition*,
Intelligent Systems Reference Library 80, DOI 10.1007/978-3-662-45704-7_2

genetic modifications in agriculture has the potential to strengthen the economies of industrialized countries, lower pesticide use, and combat hunger crises in developing countries. Others have resisted the spread and implementation of these biotechnology applications. Concerns have focused on the capacity of genetically modified foods to cross biological boundaries, causing harm to humans and the environment. However, resistance also stems from the post-material values movement of the 1960s and 1970s that highlighted negative sociological externalities of biotechnology, including the commodification of life and the increase of inequality [4].

Policymakers in industrialized countries have responded differently to these perceived opportunities and threats. A biotechnology policy divide has emerged since the 1990s between North America and the European Union (EU), while South American countries like Brazil have pursued an inconsistent policy trajectory [5–13]. The influences of socioeconomic conditions, political institutions, informal and formal participants in public policy decision-making, the media, and especially the contrasting policy implications of the "process" and "product" approaches to biotechnology regulations embraced by the US, Canada, Brazil, and EU are often cited to explain differences in policy design and implementation. This study, which highlights political culture and risk perceptions as special to understanding the complexity that characterizes this policy divide and policy inconsistencies, seeks to enhance our understanding of the remarkably different approaches taken by policymakers cross-nationally.

Do national differences in political culture, as expressed through different levels of tolerance for uncertainty and risk, affect the formulation of protective regulatory policy in the area of genetically modified food? Using consumer survey data and a detailed examination of the regulatory policies pursued in different national contexts, the study hypothesizes that varying levels of uncertainty tolerance coupled with prevailing risk perceptions either encourage the implementation of new protective policies or lead to the adjustment of existing regulations. This study applies elements of the cultural model developed by Hofstede [14] and uses a modified version of the *Margolis Risk Matrix* [15] to assess risk tolerance in different national contexts. Following a brief review of the literature about the influence of political culture and risk perceptions on policymaking, the paper compares the development of genetically modified food policy in the United States, Canada, Brazil, and European Union.

2.2 Political Culture, Risk Perceptions, and Policymaking

Discernible values and political cultures within and across countries shape citizen interactions with governments and influence policy processes. In the United States, researchers have identified a number of "major value orientations" and political cultures by region [16, 17]. Values such as individual freedom, equality, and progress, coupled with an individualistic, moralistic, and traditionalistic political culture, have implications for policymaking. For several decades, the meaning and significance of political culture for the functioning of a democratic government has been an integral part of the scholarly discourse in political science [18–21]. Despite the volume of research that regards political culture as an important contextual variable, the intersection of political culture and policy processes has required researchers to go beyond the traditional political culture literature.

Work in the area of international management links different aspects of political culture within countries to both the operation of economic organizations and the unfolding of political processes [22, 23]. In line with Montesquieu's notion of the general spirit of a nation, Hofstede argues that the unique characteristics of political institutions, governmental arrangements, laws, and legal systems are the tangible manifestations of differences in the national identity or political culture of a given country or geopolitical region [24, 25]. Visible to the observer, these differences in political culture can be discerned, measured, and quantified into indexes applicable across different countries. Operationally, political culture may be assessed along a number of interrelated dimensions, including power distance, collectivism versus individualism, femininity versus masculinity, long- versus short-term orientation, and uncertainty avoidance.

As suggested by Hofstede, societies deal differently with ambiguities or uncertainties that are the result of advances in technology. Depending on how much uncertainty a society can tolerate, the degree of rejection or acceptance of new products by society and the corresponding legal and regulatory regime discussed and implemented by governments differ. Especially useful here is the uncertainty avoidance index developed by Hofstede, which is inversely related to the acceptance of new products [26]. The index reflects the extent to which members of a society attempt to cope with anxiety by minimizing uncertainty. The researchers provide a useful analytical tool to establish a link between political culture and policy processes. Considering several interrelated broadly conceived core cultural dimensions (e.g. power distance, collectivism, individualism, femininity, masculinity, and uncertainty avoidance) that can be reasonably generalized across countries and regions, the assessment tool developed by Hofstede offers a sound approach to understanding the influence of political culture on policy processes.

Along the lines of Charles-Louis de Montesquieu's notion of the general spirit of a nation, the researchers argue that the diversity of political institutions, government, laws, and legal systems, and so on are the manifestations of differences in the national identity or political culture of a given country. One of the critical aspects of political culture that influence policy processes is the way societies deal with ambiguities or uncertainties. While advances in technology can reduce uncertainties, the unknown health and environmental effects of new technologies, like the genetic engineering of food, can nourish uncertainties within societies. Depending on how much uncertainty a society can tolerate, governments discuss and implement different kinds of laws to deal with and reduce uncertainties. Within the broader context of different national identities in terms of their essential patterns of thinking and the subsequent emphasis of values, symbols, and rituals, Hofstede identifies five dimensions of political culture (i.e. power distance, collectivism vs.

individualism, femininity vs. masculinity, long vs. short term orientation, and uncertainty avoidance), and develops indices for each dimension usually ranging from 0 to 100 based on extensive survey research conducted in more than fifty countries.

The uncertainty avoidance index reflects the extent to which members of a society attempt to cope with anxiety by minimizing uncertainty, which is not to be confused with risk avoidance. Given the tensions between threats from the unknown and the need for predictability, the uncertainty avoidance index suggests that, in contrast to countries or regions characterized by high levels of uncertainty tolerance (e.g. Southern Europe and Latin America), societies characterized by low levels of uncertainty tolerance (e.g. Scandinavian countries and Northern America), are less confident in their ability to influence government and tend to prefer structured circumstances expressed by "more and more precise laws" [25].

In addition to the level of uncertainty tolerance among citizens of a given country, or countries across a region, different risk perceptions among policy stakeholders influence policy processes. The complexity of the policy environment in which considerations of risk arise as well as perennial confusion over how to use the concept of risk in practice compound the lack of clear information about risk. Risk is the "down side of a gamble ... [which] implies a probability of outcome, and the gamble may be involuntary or voluntary, avoidable or unavoidable, controllable or uncontrollable. The total gamble in which the risk is embedded must be addressed if the risk is to be analyzed, both the upside (benefits) and down side" [27]. Thus, a risk is fairly straightforward, yet assessing its impact within a policy debate is difficult because of competing claims, issues and interests [28].

Perceptions of and predispositions toward risk are based on patterns of thinking, or mental models, [29] which can be defined as personal constructs that vary by individual and constitute a complex set of perceptions, opinions, attitudes, and beliefs used to make sense of reality [30–32]. Differences in these mental models can be noticeable, as they may affect both decision-making processes and their consequences [33, 15]. Non-expert lay observers outside the scientific community (i.e., the public) tend to rely on cognitive heuristics in their approach to assessing health and environmental risks more than experts within the scientific community [34–37]. Scholars have also considered the negative consequences that arise in the context of the expert-lay person dichotomy and have developed different models of risk perception. As a consequence, ethical concerns expressed by the public regarding major technological advances have been all but ignored by expert institutions [38]. Others argue that in addition to traditional factors like novelty and dread, concerns about "interference with nature" play a major role in accounting for the perceived risk of genetic engineering [33].

Margolis also provides a useful analytical framework for examining the influence of different risk perceptions on policymaking in the areas of health and the environment. The difference in attitude between experts and the lay public create rival mental models that affect both the choice of policy solutions and the solutions available. It is these different judgmental heuristics that create both consistencies and variation in risk evaluation. Thus, experts and the public may experience different decisional dilemmas and varying risk perceptions. Perceptions concerning the dangers and opportunities of a given situation may lead to differences in the scope of regulatory approaches by government. To differentiate rival risk perceptions, the *Margolis Risk Matrix* suggests that an individual interprets a situation as: one that creates opportunities; one that presents threats; one that contains both opportunities and threats; or, as one that offers neither opportunity nor threat. Along the lines of any tangible costs and benefits (or dangers and opportunities), the *Margolis Risk Matrix* proposes distinct risk perceptions that can be applied to the general public and policymakers. These stakeholders often seek and rely on expert advice [15].

The specific types of risk perceptions that guide the decision-making process include: fungibility or balanced risk taking (seeing both dangers and opportunities); cautious or "better safe than sorry" risk aversion (seeing dangers but no opportunities); opportunistic or "waste not, want not" risk taking (seeing no dangers but opportunities); and, indifference or "move along, go along" risk indifference (seeing no dangers and no opportunities). The balanced risk position suggests that individuals who are aware of the dangers act to somehow trade off potential benefits. Persons who are guided in their assessment by the indifference risk position see neither dangers nor benefits and, as such, a given policy issue is off-screen and no response is to be expected. Finally, the cautious and opportunistic risk positions suggest that either dangers or benefits—but not both—guide a person's risk assessment and response to a policy issue. The combined use of the uncertainty avoidance index and the *Margolis Risk Matrix* as an analytical framework measures uncertainty tolerance across countries and regions and the prevailing risk perceptions among the relevant policy stakeholders [15].

2.3 Research Design

Relying on both the uncertainty avoidance index to understand the national or regional context and the *Margolis Risk Matrix* to assess the risk perceptions among the policymakers and the public, this study traces the policy trajectories of genetically modified food regulations in the United States, Canada, Brazil, and within the European Union between 1990 and 2006. The study hypothesizes that low levels of uncertainty tolerance and the prevalence of reasonable risk taking coupled with cautious risk perceptions encourages the formulation of stringent protective regulatory policies. On the other hand, high levels of uncertainty tolerance and the prevalence of indifference coupled with opportunistic risk perceptions among policy stakeholders encourages the continuation or adjustment of existing protective regulatory policies. Finally, regardless of low or high levels of uncertainty tolerance, the simultaneous and equally strong competition of cautious and opportunistic risk perception facilitate the development of an inconsistent protective regulatory framework.

In light of these research expectations, it is important to distinguish normal public policy from protective public policy. According to James Anderson, public policies consist of a "purposive course of action followed by an actor or set of actors in dealing with a matter of concerns." Public policies, he makes clear, are those laws and regulations "developed by governmental bodies and officials" [39]. Accordingly, protective regulatory policy, a type of policy output that is often associated with environmental regulations at the national level, is defined as a purposive action by government to enhance, protect, or maintain public health and safety in response to actual or potential hazards or threats that originate within the private sector [40–42].

In this study, the outcome of interest is the adoption of new or modification of existing protective regulatory policy in the area of genetically modified food. The uncertainty tolerance level and risk perceptions among policy stakeholders in a particular country or region are used to predict the appearance of new protective regulatory policies. Given the focus on genetically modified food, policy stakeholder representation is limited to the major regulatory policy institutions and scientific advisory committees dealing with genetically modified foods in the US, Canada, Brazil, and EU. They include the Food and Drug Administration and National Research Council in the United States, Health Canada, the Canadian National Biotechnology Advisory Committee, and the Royal Society of Canada. In regards to Brazil the focus rests on the National Biosafety Technical Commission, while the European Parliament, Commission, and Council of Ministers serve as the primary regulatory EU institutions.

The uncertainty tolerance level, defined as "the extent to which the members of a culture feel threatened by ambiguous or unknown situations," [25] is measured by the uncertainty avoidance index developed by Hofstede (see Table 2.1 for uncertainty avoidance index rankings and scores by country and region). Based on scores derived from survey research, the uncertainty avoidance index captures variations of risk avoidance attitudes across different countries and regions and provides the overall context for different policy related outcomes. It does not capture changes in uncertainty values over time nor negative attitudes towards a particular technology. For the present study, risk is defined as the assessment of the threats and opportunities presented by a potentially hazardous situation.

The uncertainty index is constructed using the country mean scores for the following three questions: (1) *Rule orientation*. Agreement with the statement: "Company rules should not be broken—even when the employee thinks it is in the company's best interest"; (2) *Employment stability*. Whether employed respondents intend to continue with their current employer either for 2 years or less, or from 2 to 5 years; and, (3) *Stress*. Expressed in the answer to the question: "How often do you feel nervous or tense at work?" The index normally has a value between 0 (weak uncertainty avoidance) and 100 (strong uncertainty avoidance).

The United States, which has an uncertainty avoidance index score of 46 out of 112 and is ranked 43 out of 50 countries and 3 regions, is characterized by high levels of uncertainty tolerance about new technologies (see Table 2.1). Canada ranks 41 with an uncertainty avoidance index score at 48. As such, Canada, which

Table 2.1 Uncertainty avoidance index (UAI) values for 50 countries and 3 regions

Score rank	Country or region	UAI score	Score rank	Country or region	UAI score
1	Greece	112	28	Equador	67
2	Portugal	104	29	Germany FR	65
3	Guatemala	101	30	Thailand	64
4	Uruguay	100	31/32	Iran	59
5/6	Belgium	94	31/32	Finland	59
5/6	Salvador	94	33	Switzerland	58
7	Japan	92	34	West Africa	54
8	Yugoslavia	88	35	Netherlands	53
9	Peru	87	36	East Africa	52
10/15	France	86	37	Australia	51
10/15	Chile	86	38	Norway	50
10/15	Spain	86	39/40	South Africa	49
10/15	Costa Rica	86	39/40	New Zealand	49
10/15	Panama	86	41/42	Indonesia	48
10/15	Argentina	86	41/42	Canada	48
16/17	Turkey	85	43	USA	46
16/17	South Korea	85	44	Philippines	44
18	Mexico	82	45	India	40
19	Israel	81	46	Malaysia	36
20	Colombia	80	47/48	Great Britain	35
21/22	Venezuela	76	47/48	Ireland (Republic of)	35
21/22	Brazil	76	49/50	Hong Kong	29
23	Italy	75	49/50	Sweden	29
24/25	Pakistan	70	51	Denmark	23
24/25	Austria	70	52	Jamaica	13
26	Taiwan	69	53	Singapore	8
27	Arab countries	68			

Source Hofstede [24], p. 113

serves as a control case in comparison to its southern neighbor, ranks slightly stronger in terms of uncertainty avoidance than the United States. In contrast to the United States and Canada, South American countries like Brazil are generally characterized by higher levels of uncertainty tolerance. Based on an index score of 76, Brazil ranks 21/22. Turning to Europe, an overwhelming majority of EU member states, including Germany, France, Italy, and Spain, with respective uncertainty avoidance index scores of 65, 86, 75, and 86, rank much higher on the uncertainty avoidance index. Similar to Brazil, this suggests low levels of uncertainty tolerance—and thus, presumably, a desire for more stringent regulatory policies compared to their North American counterparts. With the notable exception

of the United Kingdom, the low uncertainty tolerance countries include the dom-
inant policy actors within the EU. Based on a combined average, Germany, France,
Italy, and Spain rank 18 with an average uncertainty avoidance index score at 78.

For the analysis, the prevailing risk perceptions in a given country or region are
identified along the balanced, cautious, opportunistic, and indifferent risk trajectories. Within the balanced risk position, stakeholders perceive risk in terms of high
threat for the well being of society or the individual but also high opportunity for
gaining tangible benefits. Public opinion and official policy statements or actions
that present trade-offs between these threats to the well being of society and
socioeconomic benefits illustrate the balanced risk perception.

From a cautious risk perspective, stakeholders perceive risk in terms of high
threat to the well being of society or the individual and low opportunity for gaining
tangible benefits. Public opinion and official policy statements or actions that
strongly emphasize threats to the well being of society relative to socioeconomic
benefits illustrate the cautious position. An opportunistic risk perception is characterized by low threat to the well being of society or the individual and high
opportunity for gaining tangible benefits. Public opinion and official policy statements or actions that overemphasize socioeconomic benefits relative to threats
illustrate the opportunistic risk assessment. Finally, stakeholders guided by indifference perceive risk in terms of low threat to the well being of society or the
individual and low opportunity for gaining tangible benefits. Public opinion and
official policy statements or actions that neither emphasize threats nor socioeconomic benefits to society illustrate the indifferent risk perception.

Poll results and document analysis of official policy statements, reports, and
regulations were analyzed to assess the respective risk perceptions of policy
stakeholders in the United States, Canada, Brazil, and European Union. Admittedly,
data derived from document analysis alone has shortcomings. It is not possible, nor
does this study claim, to trace regulators' thought processes. Rather, the evidence
here examines the prevailing risk perceptions over time. Given the inconsistent
availability of opinion polls regarding the genetic modification of food between
1990 and 2006, this study relies on different surveys and opinion polls conducted
by research organizations and academic institutions. Similar question wording
regarding the public's attitudes towards genetically modified food across different
survey administrations allow for a comparison of risk perceptions across countries
over time.

For the United States, the sampling period extends from 1990 to 2006. Relevant
opinion surveys include the 2001–2006 Pew Initiative on Food and Biotechnology,
a 1999–2000 Gallup Poll, and a wide variety of other studies conducted by research
organizations and policy institutes such as the Food Policy Institute at Rutgers
University [43–55]. For the EU, the 1991–2005 Eurobarometer surveys capture the
attitudes regarding the risk perceptions associated with genetically modified foods,
while public opinion polls conducted by a number of Canadian academics and
research organizations (e.g., Decima Research and Pollara Research) illustrate the
relevant Canadian attitudes between 1997 and 2006 [56–59]. While there is an
extensive array of opinion polls available for the United States and the EU, the

range of surveys to understand the attitudes towards genetically modified foods in Brazil remains somewhat limited. Nevertheless, the scientifically-based public opinion polls sponsored by the Brazilian Institute of Public and Statistical Opinion (IBOPE) as well as those conducted by scholars in regards to specific segments of society allow for accurately capturing the relevant attitudes between 2001 and 2006 [60–64].

2.4 Risk Perceptions and Policy Trajectories

2.4.1 The United States

The existence of genetically modified (GM) food in the United States became widely known with the approval of recombinant bovine growth hormones in 1993, the commercialization of the first genetically engineered tomato in 1994, and the approval of other genetically engineered products like cotton, soybeans, and squash by 1996. Based on a pro-business and anti-regulatory consensus pursued in tandem by the United States government and the influential biotechnology industry, the relevant regulatory framework was well established by the 1980s and reflected the "optimism about progress in the natural sciences and related technological innovations on the conviction that society would benefit more from GM technology if governments would interfere as little as possible and avoid the introduction of specific legislation" [65]. Following the regulatory adjustments proposed by the President's Office of Science and Technology Policy and enshrined in the 1986 *Coordinated Framework for Regulation of Biotechnology,* the Food and Drug Administration in its 1992 statement of policy, *Foods Derived from New Plant Varieties,* reiterated the product-based approach of the regulatory regime. Accordingly, genetically modified foods are held to pose no safety concerns because "many of the food crops currently being developed with gene splicing techniques do not contain substances that are significantly different from substances already in the diet" [66].

Within the context of emphasizing the safety or minimal dangers to human health and the environment of these modified foods, minimize regulatory burden, and facilitate the development and commercialization of such products, public opinion and especially official statements associated with the Food and Drug Administration highlighted the tangible spillover benefits of genetically modified food for society. The public, largely unaware of the major technological changes in agricultural food production, uninformed regarding the potential negative environmental effects of genetically modified food, and largely excluded from the relevant decision-making processes that ultimately determined the commercial marketing of genetically modified food, had little basis for assessing the potential dangers of engineered food. Within this broader context of low awareness and a regulatory approach that limited public input, concerns regarding genetically modified food among the public were not well organized or given much credence. While the public knew little about biotechnology applications in general, perception of genetically modified food in the context of biotechnology applications was generally positive and emphasized consumer benefits of such applications [51, 55, 67, 54]. Support rates consistently hovered around 70 % during the 1990s, illustrating both considerable support and "remarkable stability of people's opinions on biotechnology in the US" [68].

This positive public opinion embedded within a utilitarian worldview of technological advances and coupled with the government's strong support for scientific research on food genetics as well as the courts' positive assessment of biotechnology regulations framed the oversight functions at the agency level [69]. Staffed with many former employees of major agribusiness corporations, the Food and Drug Administration cooperated closely with entities like Monsanto and touted the benefits of GM food, as illustrated by the approval of the recombinant bovine growth hormone in 1993 [70–74]. As a consequence of this mutually opportunistic risk perception among regulators and agribusiness representatives, public statements by officials within the agency emphasized that genetic engineering of food would contribute to "enhanced resistance to disease, pests and herbicide in major field crops. For biotechnology techniques applied to feed grain and forage crop production, consumer effects will almost exclusively be cost reduction" [75].

In light of these favorable claims, and the concerted lobbying efforts by agribusiness, [76] the Food and Drug Administration's Center for Food Safety and Applied Nutrition generally opposed labeling requirements for genetically modified food unless their nutritional content was substantially modified. The implementation of a mandatory labeling requirement would "increase the cost of these foods to consumers and would disrupt our complex food distribution system" [77]. Although the Food and Drug Administration provided guidance to the industry as to how they may voluntarily label genetically modified foods, the agency also maintained that "bioengineered foods [do not] differ from other foods in any meaningful or uniform manner, and that GM foods as a category of food products do not present any different or greater safety concern than foods developed by traditional plant breeding" [69].

That genetically modified food was considered unlikely to pose any hazardous risk to the public health became apparent during the Food and Drug Administration's approval of numerous genetically modified products between 1994 and 2007 [78]. The Flavr-Savr tomato offers a case in point. Developed by Calgene, a small company based in California that in 1996 was taken over by Monsanto, [79] the Flavr-Savr was subjected to a comprehensive approval process by the Food and Drug Administration. In its document on *Foods Derived From New Plant Varieties* and other public statements, the agency viewed the genetically modified tomato as beneficial to the consumer and deemed it to pose no environmental risks [66, 80, 81]. Genetically modified foods like the Flavr-Savr were characterized by "improved shelf-life, processing characteristics, flavor, nutritional properties, and agronomic characteristics, such as tolerance to chemical herbicides and resistance to pests and disease" [76].

Although the genetically modified tomato was eventually taken off the market in 1997 due to poor yield in the unsuitable sandy soil and humid climate of Florida, the Food and Drug Administration stated during the initial approval process that "the intended effect of the altered RNA of the new PG (polygalacturonase) gene that suppresses the breakdown of pectin in Flavr-Savr tomatoes does not raise safety questions. Pectin is a part of many fruits and is generally recognized as safe (GRAS) substance" [80].

As the approval process of genetically modified food developed between 1994 and 2007, the public continued to associate genetically modified food with mostly low threats to human health and saw the possibility of gaining benefits from it. Poll results from the mid- to late-1990s seemed to confirm the public's positive attitudes towards genetically modified food. Assuming that engineered food would improve the quality of life and benefit society, a majority of the public continued to believe that tangible gains could be derived from genetically modified food [46, 82]. However, the formation of the Organic Consumer Association in the late 1990s, the anti-GM food campaigns organized by voters to require mandatory labeling during the early 2000s, and surveys conducted by the Pew Initiative on Food and Biotechnology between 2001 and 2006 also illustrate a shift in public opinion characterized by the emergence of a visibly cautious risk perception mixed with elements of an opportunistic risk position.

In contrast to the 1990s, when public attitudes were generally supportive of genetically modified food and few consumer interest groups considered potential biotechnology threats to be a high priority, a much more skeptical public has emerged over the past decade. Survey results from the Pew Initiative on Food and Biotechnology have since 2001 indicated that a relatively small segment of society, around 25 %, expresses outright "support for genetically modified foods" [83]. Parallel to this mixture of opportunistic and cautious risk perceptions, low knowledge and awareness about food biotechnology applications continued to play a major role in public opinion polls, which through 2006 showed that a majority of the public had not heard much about biotechnology or knew very little about the various biotechnology applications [46, 49, 51, 84, 85]. According to Mark Winston, a close observer of the biotechnology debate, the public "has been besieged by sound bites and public relations hype rather than exposed to comprehensive and informed debate and dialogue" [86].

Within this broader context, Americans remained confident in the ability of the appropriate regulatory agencies to guarantee the introduction of safe biotechnology products and ensure the maintenance of public health [87]. As these public perceptions evolved, the Food and Drug Administration continued to emphasize the low threats and benefits of genetically modified food by referring to the "substantial equivalence" principle (i.e., the undistinguishable nature of genetically modified food from conventional food). Accordingly, a 1995 policy statement by the Food and Drug Administration stipulated that no formal review was needed for engineered food:

Based upon the extensive history of safety of plant varieties developed through agricultural research, the Food and Drug Administration has not found it necessary to review the safety of food derived from new plant varieties. [Moreover] the Food and Drug Administration is not aware of information that would distinguish genetically engineered food as a class from food developed through other methods of plant breeding [88].

The belief that biotechnology "greatly expands the pool of potentially useful traits available" and the minimal concerns regarding allergic reaction and antibiotic resistance characterized the agency's fundamental perspectives on genetically modified food as both beneficial and safe [89].

The StarLink corn saga that played out between 1997 and 2001 shook public confidence in the food manufacturing industry but this crisis "did not lead to a visible consumer reaction, like the shoppers panic that would surely have occurred in Europe" [71]. Press reports and public statements from Friends of the Earth suggested a widespread "commingling" of StarLink, a genetically engineered corn plant with the ability to encode the Bt protein Cry9c that was not approved for human consumption, with non-genetically modified corn destined for human consumption. Tests confirmed by the Food and Drug Administration in 2000 found StarLink traces in taco shells [90]. Despite these events and the recall of various foods by producers in response to the Food and Drug Administration's continued StarLink investigation, the agency's approach to regulating genetically modified food did not change during this period. Not only did the agency maintain its 1994 policy of voluntary consultation with the biotechnology industry to assist in the safety assessment of genetically modified products entering the food chain, the Food and Drug Administration also continued to stress that genetically engineered food was safe and beneficial.

Agricultural research has shown that "most of the substances that are being introduced into food by genetic modification have been safely consumed as food [already] or are substantially similar to such substances" [88]. The Food and Drug Administration continued to emphasize the safety of genetically modified food, as illustrated by James Maryanski's testimony before the Senate Committee on Agriculture, Nutrition and Forestry in the fall of 1999. Maryanski, the agency's biotechnology coordinator, stated that, "In most cases, these genes [recombinant DNA] produce proteins, or proteins that modify fatty acids or carbohydrates in the plant, in other words, common food substances" [91]. Before a Senate hearing a year later, Joseph A Levitt, the Food and Drug Administration's director of Food Safety and Applied Nutrition, echoed these sentiments of no known dangers [92]. The agency continued to stress the benefits associated with food biotechnology, including the reduction of chemical pesticides and herbicides and the possible improvement of food's nutritional properties [93].

The regulatory changes proposed since 2000 by the Food and Drug Administration, coupled with the agency's long-standing awareness that the introduction of genetically modified proteins into food may cause allergic or toxic reactions in consumers, did not challenge the prevailing opportunistic risk perception within the agency. In response to public concerns, the Food and Drug Administration in May 2000 proposed changing the voluntary evaluation or consultation procedures that guided the pre-market notification program for genetically modified food. Until then, food companies were not required to seek pre-market consultation on new genetically modified products.

Officially announced on January 18, 2001 in conjunction with a proposal for voluntary labeling of genetically engineered food, the adjusted consultation rule required genetically modified food developers to submit data regarding plant-derived genetically engineered food at least 120 days before releasing an engineered food product into the market [94]. This mandatory pre-market notification proposal, later complemented by guidance on the evaluation of genetically modified plants intended for food use and posting of the consultation results on the Food and Drug Administration website, appeared to represent a fundamental change in the agency's risk perception [95]. However, while the Food and Drug Administration has continued to make the consultation results available online, it dropped the mandatory pre-market notification and voluntary consultation plan in 2003 and reiterated that transferred genetic materials do not pose any significant safety concerns [94–98].

Guided by an opportunistic risk perception, the Food and Drug Administration remained firm on the issue of mandatory genetically modified food labeling. The agency strongly believed that food created through biotechnology was identical to food developed using conventional plant breeding methods. Thus, while the Food and Drug Administration agreed to voluntary labeling, as suggested by the 2001 document, *Draft Guidance for Industry: Voluntary Labeling Indicating Whether Foods Have or Have Not Been Developed Using Bioengineering*, [99] it did not require any special labeling to distinguish GM food from non-GM food [96]. Policymakers have not changed their views on labeling or their risk assessment of genetically modified food despite mounting pressures, namely: consumer concerns and demands for the right to know which foods have been genetically engineered; repeated introduction of a bill in Congress to require genetically modified labeling, known as the *Genetically Engineered Food Right-to-Know* Act (H.R. 5269 2006); the enactment or serious consideration of food labeling regulations at the state level; and, recent food scares such as the ProdiGene affair.

The ProdiGene affair, which involved field trials of pharmaceutical maize conducted by Texas-based ProdiGene to produce a vaccine that prevents diarrhea in pigs influenced but did not change prevailing risk perceptions within the Food and Drug Administration. Ultimately, the agency decided to order the destruction of fields in Nebraska and Iowa that were contaminated with genetically modified corn. Since the end of 2002, the agency has proposed strategies to minimize the inadvertent introduction of genetically modified materials into agriculture, the environment, and the food supply. Despite other incidents (e.g., the Ventria affair, in which California-based Ventria Bioscience developed transgenic rice varieties to be openly grown in trial plots located in the rice growing area of California's Central Valley), the Food and Drug Administration's emphasis on the minimal dangers and discussion of the tangible benefits of food biotechnology suggest the opportunistic risk perception within the agency continues to prevail [100–102].

2.4.2 Canada

The U.S. proposal to release engineered organisms for field testing sparked a debate in Canada about biotechnology application in the early 1980s. Policymakers and national advisory committees showed strong support for biotechnology. Published by the Canadian Ministry of State for Science and Technology in 1980, *Biotechnology in Canada* laid the policy groundwork for the "promotion and development of biotechnology" and the establishment of a private-sector task force on biotechnology [103]. With the aim of accelerating commercial progress and maintaining competitiveness relative to other countries in biotechnology research, the Canadian government invested millions of dollars to institutionalize the 1983 *National Biotechnology Strategy* and fund national biotechnology research centers. This in turn led to the establishment of the National Biotechnology Advisory Committee, whose members—drawn from academia, the private sector, and government—were charged with providing advice to the Science and Technology ministry on a national biotechnology strategy.

In 1984, the first report published by the National Biotechnology Advisory Committee foresaw an active role for the federal government in shaping biotechnology policing and used the government to "take advantage of the current window of opportunity in biotechnology" [103]. Sensitive to domestic and international pressures to develop biotechnology, senior officials within the Canadian agriculture bureaucracy also emphasized the benefits of new biotechnologies and stressed the need to develop relevant regulations that would protect human and animal health while safeguarding the environment and promoting a competitive advantage for industry [103, 104].

Although the potential hazards of genetically modified organisms were actively debated in response to a 1989 report by the Ecological Society of America, the Canadian government did not reconsider its favorable stance on biotechnology and continued with the formulation of a relevant regulatory framework. Driven by the consensus to achieve progress through biotechnology and the increasing conviction that engineered food products were as safe as conventional products, policymakers began to lay the foundation for the 1993 *Regulatory Framework for Biotechnology*. While the document disregards most social, economic, and ethical issues raised by the new technologies, the regulatory framework coupled science-based risk assessment with other internationally recognized and established risk assessment concepts. By then the notions of *familiarity* and *substantial equivalence*, advocated by various national and international organizations, including the OECD, National Academy of Sciences in the US, United Nations' Food and Agriculture Organization, and World Health Organization, became the main regulatory principles guiding Canadian policymakers in crafting the regulation of biotechnology applications [105–107].

Guided by these principles and relying on "information and advice from scientific networks and advisory committees in developing the genetically modified policy and regulatory framework" [108]. Within a regulatory environment where participation in decision-making is exclusionary and judicious, the Canadian government avoided public and parliamentary debates and decided to divide regulatory responsibilities among Environment Canada, Health Canada, and the Canadian Food Inspection Agency. Environment Canada assumed responsibility for assessing the environmental risks of biotechnology products and Health Canada, the Canadian counterpart to the U.S. Food and Drug Administration, was charged with the regulation of genetically modified food based on B.28.001, B.28.002, and B.28.003 of the 1920 Food and Drugs Act.

Yet, despite these regulatory adjustments, there are important differences between Canada and the United States. Under Division 28 of the Food and Drug Regulations (Novel Foods), Health Canada considers any genetically modified food a *novel* food by definition and follows a formal pre-market notification policy that requires manufacturers and importers of genetically modified food to submit data to Health Canada for a pre-market assessment. Furthermore, Canada's consolidation of the food inspection service during the 1990s culminated in the establishment of the Canadian Food Inspection Agency, which is responsible for monitoring and implementing the policies of Environmental Canada and Health Canada [107, 109, 110].

That the assessment of novel food by Health Canada made use of the substantial equivalence principle became apparent with Health Canada's approval of genetically modified food based on the *Guidelines for the Safety Assessment of Novel Foods* [111]. Operating under these guidelines and in some cases hastened by major international agribusinesses like Monsanto, Health Canada has approved more than 90 novel foods since the mid-1990s, ranging from novel varieties of corn and potatoes to soybeans and tomatoes [72, 112]. As in the United States, the Flavr-Savr™ tomato made its debut in the mid-1990s and like its southern neighbor, the Canadian government did not require any labeling. Comparing the Flavr-Savr™ tomatoes to other non-genetically engineered counterparts, Health Canada "found no difference in composition or nutritional characteristics. Based on Calgene's information, the Department found the Flavr Savr to be as safe and nutritious as other tomato varieties" [113]. By acknowledging that this genetically modified product is engineered to "ripen longer on the vine than other tomatoes in order to more fully develop its flavor," Health Canada also acknowledged the benefits of the novel tomato.

As the approval of genetically modified food continued to rely on the assessment of scientists working for the government and a regulatory framework that did not provide for independent scientific review and public involvement in product assessment, no major public controversies regarding the regulatory framework and genetically modified foods emerged [105]. In fact, poll results illustrate that the public was scarcely aware of these applications. A national survey conducted in 1997 by Einsiedel and Medlock asked: "What comes to mind when you think about biotechnology in a broad sense, that is, including genetic engineering?" Only one third of respondents answered this open-ended question [59]. A second national opinion poll conducted in 1999 confirmed public unfamiliarity with biotechnology applications [58].

However, as nongovernmental organizations became increasingly visible in opposition to biotechnology applications and the news media began paying more attention to covering biotechnology events, awareness increased. A national survey conducted in 2000 revealed a significant change in the level of public awareness. In response to the same open-ended question concerning biotechnology and genetic engineering, more than 75 % of respondents ventured an answer [59]. Since 2001, overall familiarity with and support of biotechnology has steadily grown [56, 114]. At the same time, "there remains continued and widespread wariness about GM food," according to Pollara Research [114].

Stressing that genetically modified products are not inherently different from their naturally grown counterparts, the Canadian government has continued to emphasize the safety and benefits of genetically modified food as economically beneficial and innovative. The approval guidelines for novel foods, released in 1994, the *Canadian Biotechnology Strategy* published in 1998, and a report titled *Biotechnology Transforming Society* published by the Canadian Department of Foreign Affairs and International Trade in 2003 all emphasized the benefits associated with the new technology [115–117]. Similar to the United States, a scientifically rational focus embedded within an opportunistic risk perception remained the hallmark of the regulatory food biotechnology framework in Canada, despite increasing international attention to genetically modified food and domestic skepticism regarding the Canadian genetically modified food regulatory framework at the dawn of the twenty-first century. In 2000, the Codex Alimentarius Commission, formed jointly by the World Health Organization and the Food and Agriculture Organization, enumerated several universal principles regarding the safety of genetically modified food and called for explicit labeling of such food products [118].

In light of the increasing controversies around genetically modified food in Canada, the Royal Society of Canada, an independent panel of scientists, published *Elements of Precaution: Recommendations for the Regulation of Food Biotechnology* in 2001. This report pointed to significant shortcomings in the existing risk assessment procedures used by the Canadian government and concluded that the Canadian regulatory framework failed to conform to scientific standards.

Filled with more than 50 recommendations, including a call to make public experimental protocols and data, *Elements of Precaution* urged the Canadian government to broaden and strengthen the biotechnology regulatory system. The government had already begun to move in this direction, reconsidering the 1983 *Canadian Biotechnology Strategy* and establishing a new advisory body, the Canadian Biotechnology Advisory Committee, in the late 1990s. In response to the Royal Society of Canada, the government announced that it would make changes to its risk assessment procedures. However, the government's *Action Plan*, a series of progress reports published in 2001, does not indicate any fundamental regulatory changes—nor does the government's assessment of labeling as expensive and impractical [107, 119].

Though the regulatory approach remains fundamentally unchanged, public acceptance of biotechnology applications changed considerably between 1997 and

2006—a trend that has softened the opportunistic risk perception of genetically modified food among the Canadian public in favor of a more cautious approach. Although survey results show a decrease in approval regarding the acceptability of various biotechnology applications, including genetically modified food, public opinion remained generally favorable into the mid-2000s. In 1997, 67 % of the public *definitely agreed/agreed* that genetically modified food was useful; by 2000 the equivalent figures dropped to 57 for the combined categories. For the same time period, risk perceptions about genetically modified food remained high, while a decreasing but still substantial percentage of respondents remained supportive of encouraging genetically modified food applications in 2001. As indicated by focus group studies that were conducted since 2001, Canadians have become increasingly skeptical of genetically modified food with a substantial segment of society expressing the belief that specific biotechnology applications, such as engineered fish and agricultural products, will have more negative than positive effects [56, 114].

2.4.3 Brazil

A relatively orderly policy process and consistent risk perceptions underpinned the regulation of genetically modified foods in the United States and Canada. In contrast, the regulatory policy trajectory in Brazil was one characterized by paradoxes and mutually exclusive and competing risk perceptions within and at different levels of government. Brazil's food biotechnology regulatory framework can be traced to 1986. In that year, a state-owned research enterprise associated with the Ministry of Agriculture, Livestock, and Supply, the Empresa Brasileira de Pesquisa Agropecuária or Brazilian Agricultural Research Enterprise, successfully created the country's first genetically modified plant [120]. At this time, an overarching regulatory framework governing biotechnology applications did not exist. The beginnings of such a framework, profoundly influenced by Brazil's strong consumer protection movement in the wake of Brazil's democratic transition in the mid 1980s, were included in Article 225 of the 1988 Federal Constitution. It required the national government to "preserve the diversity and the integrity of the genetic patrimony of the country" and "control the production, commercialization and use of techniques, methods and substances that pose a risk to life, the quality of life and the environment" [121, 122].

Convinced by the appropriateness of the precautionary principle, the regulatory framework continued to evolve with Brazil signing the Convention on Biological Diversity at the United Nations Conference on Environment and Development in Rio de Janeiro in 1992. This Convention, designed as a practical guideline to realize the principles of Agenda 21 and signed by more than 150 governments including Canada and the EU, but not the United States, sparked Brazil's initial policy response under the Presidency of Fernando Henrique Cardoso (1995–2002). Particularly, the policy aimed at restricting both the release of genetically modified organisms into the environment and commercialization of food derived from transgenic crops. The Cardoso government, known for its implementation of market-oriented and modernizing reforms, envisioned a protective regulatory framework whereby the federal government would regulate approved biotechnology research through the Comissão Técnica Nacional de Biossegurança or National Biosafety Technical Commission. Furthermore, the administration favored a moratorium of planting of transgenic seeds for commercial purposes between 1995 and 1998; and, based on proper labeling, the commercialization of genetically modified products [13].

A major clarification of Brazil's legal and regulatory framework occurred with the passage of the 1995 Law of Biosecurity, number 8,974, which continued to rely on the EU's established norm of the precautionary principle. The law authorized the government to form a new regulatory institution, the Comissão Técnica Nacional de Biossegurança, charged with overseeing the "experimentation, registration, use, transportation, storage, commercialization, liberations, and waste removal of genetically modified materials" [13]. In order to pursue these regulatory goals within the broader institutional authority of the Ministry of Science and Technology, the 18 commission members, appointed for 2 years on a rotating basis by the Minister of Science and Technology, include representatives from the federal government, scientists, experts with scientific and technical knowledge in animal, plant, environment, and health sciences as well as civil society specialists in consumer defense and family farming. In addition to developing standards and norms in the areas of biosafety hazards and risks associated with genetically modified organisms, the actual approval process of transgenic foods requires the Comissão Técnica Nacional de Biossegurança to submit documentation to the Ministries of Agriculture, Livestock and Food Supply, Health, and Environment. Despite its diverse membership and regulatory scope, the commission has also served as a tool for policymakers in the department of agriculture to emphasize the economic and technical aspects of transgenic crops, while excluding environmental, health, and other social concerns [123, 124].

A complex web of legal and regulatory controversies unfolded following the passage of the biosecurity law. It began with the commercialization of Monsanto's roundup ready soybeans, which in reference to famous Argentine footballer Diego Armando Maradona are also known as Maradona soybeans in Brazil. In 1998, the Comissão Técnica Nacional de Biossegurança (13 votes in favor, one against, and one abstention) permitted the commercialization of Maradona soybeans. With no strings attached, this decision did neither require an environmental impact statement or labeling of the genetically modified product. Deliberating within the context of mounting pressure by Monsanto and its strategic partnership with Empresa Brasileira de Pesquisa Agropecária, the commission's technical report emphasized that "genetically modified foods do not offer risks to the environment or to health" [125]. However, this regulatory stance ignored a previous court order that ordered the pro-transgenic Ministry of Agriculture, led by agriculture minister Marcus Vincius Pratini de Moraes, to deny Monsanto the registration of roundup ready soybeans in Brazil [74, 126, 127].

A series of intense legal wrangling followed. In September 1998, a non-governmental organization and the country's most prominent consumer protection association opposed to genetic modified foods, the Instituto Brasileiro de Defesa do Consumidor or Brazilian Institute in Defense of the Consumer, argued that Maradona soybeans are substantially different than conventional soybeans. In protest to the Comissão Técnica Nacional de Biossegurança's decision, the consumer defense organization also withdrew its civil society representative from the commission and filed a lawsuit before the 6th Civil Law Circuit in Brasilia arguing that the decision ignored possible adverse effects of biotechnology on human health. Moreover, Comissão Técnica Nacional de Biossegurança's approval process violated the Federal Constitution, which required an environmental impact statement to plant genetically modified soybeans and the labeling of such products. Drawing on an international anti-genetics network and relying on banners that stated *Fankensoya: don't swallow it*, Greenpeace joined the Instituto Brasileiro de Defesa do Consumidor to challenge Comissão Técnica Nacional de Biossegurança's decision. On June 28, 2000, the Federal Court agreed with the plaintiffs and upheld the precautionary principle incorporated into the environmental provision of the Brazilian Constitution. Accordingly, the court reversed the decision handed down by the commission and required an environmental impact statement, crop segregation, and labeling [128–131].

The court decision and the justification provided by the Federal Judge, Antinio Souza Prudente, caused further controversies among the business and scientific communities. Stating that the "irresponsible spread of progress in genetic engineering would lead to damaging de-regulation of the global economy, that may at the beginning of the new millennium lead to a civilization bearing alien creatures ...," [132] this ruling, unsuccessfully appealed by the federal government and Monsanto, paved the way for a judicial moratorium on genetically modified field trials that effectively lasted until 2003. As businesses and scientists, supportive of genetically modified products, reacted strongly to the judge's anti-science word choice, the government defended transgenic foods and lauded the work of the Comissão Técnica Nacional de Biossegurança. Signed by the President's Chief of Staff and six ministers, including the Ministers of Science and Technology, the Environment, Agriculture, Justice, and Health, the government released a communiqué defending the use of genetically modified foods. It stated that the commission considered "possible risks to human and animal health and to the environment" [133] and claimed that "the government understands that Brazil cannot be outside this technology (of transgenics) or any other which might bring benefits to the country and its citizens" [134]. The Minister of Agriculture, Marcus Vincius Pratini de Moraes, accusing the non-governmental organizations of being sponsored by multi-national corporations, favored the cultivation of genetically modified foods [135]. In alliance with the Empresa Brasileira de Pesquisa Agropecária, which controlled more than 50 % of the national soybean seed production, Pratini added that "the sales of agrotoxins in Brazil could drop by 50 % with the dissemination of glyphosate-resistant GM soybeans [and] that Brazilian agriculture would be less competitive if farmers did not plant transgenic crops" [126].

Opposition to the use of genetically modified foods formed at the state level of Brazil, thereby adding another layer of competing risk perceptions. With the goal to strengthen the competitive advantage in the production of non-genetically modified crops exported to the European Union and based on a political platform that emphasized environmental protection, public health, and humans before profit, the newly elected Worker's Party government under the leadership of Olivio Dutra declared the southern state of Rio Grande do Sul free of genetically modified seeds (zona livre dos transgenicos) in 1999. Following this declaration, the State Secretary Office of Agriculture supported a ban on genetically modified crops to both strengthen soybean exports to Europe and protect public health.

These actions by a state that grew the most soy prompted 25 non-governmental organizations to form *For a Brazil Free of Transgenics*. It opposed the cultivation and commercialization of genetically altered agricultural products and argued that the effects of these products pose health risks. The efforts of *For a Brazil Free of Transgenics* resulted in legislation that outlawed the cultivation of transgenic seeds in the states of Santa Catarina, Mato Grosso do Sul, Pará, and Rio de Janeiro [136, 74].

Despite opposition at the state level and court rulings banning the commercialization of genetically modified crops, roundup ready soybeans spread rapidly across the country. Brazil's neighbor, Argentina, authorized the sale of these genetically modified soybeans in 1996. Ironically, farmers in the north of Rio Grande do Sul had been smuggling the transgenic seeds into Brazil from Argentina and illegally planting them for years. Perceived as easier to manage compared to their conventional counterpart, estimates suggested that the planting of such crops in this state increased from 15 % in 1999 to 80 % in 2004. As a result, the federal government, under its new president, Luis Inácio Lula da Silva (2003–2010), intervened to resolve the emerging legal conflicts. Officials, like federal deputy Darcisio Perondi, called for the immediate release of genetically modified soybeans arguing that "51 % of the worldwide soybean harvest is transgenic, and growing by around 20 % every year ... Transgenic crops benefit the economy and the environment as they do not require the use of agrotoxins and therefore more is produced in a smaller space and thus biodiversity is protected" [126]. By early 2003, it was also clear that over 10 % of the national 49 million tons harvest were transgenic. These facts and the government's early launch of the nationwide *Zero Hunger Program* made the destruction of such a large amount of foodstuff economically and politically unfeasible [137, 138].

In the face of these challenges and confronted by a legal environment that can be described as regulatory anarchy, the Lula administration maneuvered to take a policy stance on genetically modified foods—a difficult undertaking due to divisions and competing risk perceptions among government officials. Initially opposed to genetically modified crops, Lula acknowledged that there was "a very serious debate [about transgenic crops] within the government, because at some point we will have to say whether we are in favor or opposed. I have been strongly opposed politically today scientifically, I have doubts" [131]. Ambiguous at best, others took a much clearer stance. The Minister of Agriculture, Roberto Rodrigues, was a strong supporter of biotechnology, the Minister of the Environment, Maria Silva, an environmental activist, was opposed to transgenic food, while the head of the Brazilian Agricultural Research Enterprise, Clayton Campanhola, argued that genetically modified crops "will only be released when there is sufficient information to guarantee that there is no threat to biosecurity" [131]. Within this environment of competing and sometime ambiguous risk perceptions and faced with increasing pressure projected by organized farmers like the Federation of Farmers of Rio Grande do Sul, President Lula, despite the opposition from within government and environmental advocates, sent provisional measure 113 to Congress. Approved by Congress as law 10688 in April 2003, the measure permitted the sale of genetically modified soybeans until January 2004 and required these crops to be segregated and labeled [139–141].

Another wave of presidential decrees and provisional measures followed in order to address the legal conflicts created by the illegal planting and sale of genetically modified soybeans. Although they ultimately legalized the planting and sale of Maradona soybeans, these ad hoc and often temporary measures did not address the structural underpinnings of the existing biotechnology regulatory framework. In an attempt to do so, the new Law of Biosecurity (number 11,105), passed by Congress on March 2, 2005, revoked the 1995 Law of Biosecurity and all of the previous provisional measures. The law authorized the newly created National Council on Biosecurity under the Office of the President to formulate and implement a national biosafety policy as well as question decisions made by the Comissão Técnica Nacional de Biossegurança.

In contrast to the 1995 law, however, the reworked commission, which operates independently from the National Council on Biosecurity and whose membership increased from 18 to 27, served as the sole decision maker to approve the commercial release of transgenic organisms. More powerful than ever, the Comissão Técnica Nacional de Biossegurança became an object for both pro- and anti-transgenic camps. While tensions and polarization characterized the decision making process, the Comissão Técnica Nacional de Biossegurança, continued to weigh the risks of transgenic foods and, as of July 2011, has approved 31 such products [126, 142–144].

Parallel to the wide range of inconsistencies that surrounded the initial implementation and subsequent reorganization of the regulatory framework of genetically modified organisms, public opinion revealed consistent patterns. National surveys conducted by the Brazilian Institute of Public and Statistical Opinion in 2001, 2002, and 2003 showed that a clear majority preferred non-transgenic food over their conventionally grown counterparts. More than 70 % preferred the former, while about 15 % favored genetically modified crops. As the public became considerably more aware of such crops between 2001 and 2003, Brazilians remained skeptical about them. In fact, a substantial majority associated specific risks with genetically modified food or rejected the planting of transgenic crops until the potential risks associated with them are better understood. Accordingly, more than half were concerned that genetically modified food could damage the environment and nearly two-thirds thought that such food could pose a threat to human health. Consistent with this highly skeptical assessment, more than 70 % opposed planting of genetically modified crops [60–62]. Some of these patterns have also been confirmed by other studies. Focusing on the attitudes of young Brazilians between the ages of 16 and 24, Massarani and de Castro Moreira found that 66 % perceived biotechnology in food as socially useful. At the same time, 78 % expressed a strong sense of risks associated with this technology [63, 64].

2.4.4 The European Union

As government agencies, agribusinesses, and biotechnology firms in the United States and Canada proceeded with the commercialization of genetically engineered food and the regulatory framework for food biotechnology took different turns in Brazil, the policymakers within the EU, viewing biotechnology applications as a novel process, responded very differently. The establishment of the European Parliament's Committee on Energy, Research and Technology, and the subsequent release of the 1987 *Viehoff Report* concerning the risks of biotechnology, signaled a landmark regulatory decision. In an effort to establish a uniform regulatory approach across the member states regarding the anticipated release of genetically modified organisms, the report recommended a risk assessment of genetically engineered microorganisms and demanded a moratorium on the environmental release of such organisms "until binding Community safety directives have been drawn up" [8].

In response, the primary policy organs responsible for establishing the appropriate framework for the EU, including the European Commission and the Council of Ministers, turned their attention to the benefits and risks of biotechnology applications. In contrast to the 1976 U.S. National Institutes of Health guidelines and the favorable OECD's report on *Recombinant DNA Safety Considerations*, [145] many European policymakers, who associated genetically modified organisms with social, environmental, and economic threats, adopted an increasingly skeptical standpoint towards the unregulated application of biotechnology. They advocated the precautionary principle—a principle whose regulatory origin or gradual incorporation into the EU environmental regulatory framework can be traced to the 1969 *Swedish Environmental Protection Act* and Germany's advocacy of the *Vorsorgungsprinzip*, or cautionary principle [146–148].

Concerned about the potential risks of biotechnology and the need to safeguard the environment, the European Commission pointed to the biotechnology industries' "lack of candor … about the potential environmental risks from their products …" [71] and emphasized that "the widespread use and release of novel GMOs [genetically modified organisms] could upset the delicate balance existing in nature or even have evolutionary impacts" [146]. To avoid potentially irreversible and adverse effects of genetically modified organisms on human health and the environment, and to harmonize the national rules on the marketing of genetically modified products, a series of directives were proposed in 1994, including Council Directives 90/219/EEC and 90/220/EEC (both implemented by the Director-General for Environment and later revised by the Council Directives 94/51/EC and 94/15/EC). These directives, which were composed of more than 20 articles and were concerned with the contained use and deliberate release of GM microorganisms into the environment, reaffirmed the Commission's precautionary principle [149, 150].

Directive 90/220/EEC, for instance, cites the potential *irreversible* environmental effects of food biotechnology applications and establishes an elaborate regulatory system of placing GM products on the market. The evaluation and authorization of such biotechnology applications rely on a complex system of assessment reports and interstate information exchanges in the forms of dossiers and opinions circulated to all EU member states by the member state's appropriate Competent Authority, which is responsible for transposing directives into national law on behalf of the member state. Moreover, in case of justifiable risk, the safeguard clause under Article 16 of Directive 90/220/EEC authorizes member states to unilaterally prohibit the distribution of biotechnology products within their respective territories. The safeguard clause has been invoked on several occasions by several countries with relatively high uncertainty avoidance index scores, including Austria (three times), France (two times), and once by Germany, Luxembourg, and Greece [151].

The European Parliament in particular followed a precautionary policy approach regarding food biotechnology that reflects Europeans' strong social and cultural connection to food and their subsequent view of genetically engineered food as artificial and unnatural [152–154]. Given the logical implications of this connection within an overall environment of low uncertainty tolerance levels and the extensive publicity given in many European countries to the potential risks of genetically modified foods, the cautious risk perception of the European public has emphasized the dangers and mostly dismissed the potential benefits associated with genetically modified products. Eurobarometer surveys from 1991 and 1993 provide insight into the public's general attitudes toward genetic engineering and its different applications. Based on averages ranging from +2 (maximal support) to −2 (minimal support), no country in the EU was highly supportive of genetically modified food. In fact, support for genetically modified food remained weak at +0.47 in 1991 and +0.40 in 1993.

Within this context, and faced with public pressure throughout the early 1990s, the EU continued to closely regulate genetically modified food. Initially, the European Commission, charged with proposing legislation and overseeing the implementation of policy, favored a simple notification procedure for authorizing genetically engineered food. The Environment Committee of the European Parliament disagreed and proposed a series of amendments, requiring the labeling of genetically modified food products in 1993. While the Council of Ministers, responsible for passing EU laws, did not fully support the idea of labeling, the full plenary of the European Parliament and several member states did. The policy debate on labeling reached its regulatory apex shortly before the BSE (Bovine Spongiform Encephalopathy) or mad cow disease outbreak in the UK in January 1992, which sent shock waves throughout Europe. On March 12, 1996 the European Parliament mandated genetically modified food labeling requirements and was supported by the European Commissioner for Health and Consumer Protection [155, 156].

The controversy over genetically modified food intensified in several European countries in 1996, which was a watershed year in Europe [157]. In that year, the EU granted Monsanto to market its herbicide-tolerant soybeans and a year later Syngenta (then known as Ciba-Geigy) received permission to commercialize its insect tolerant Bt 176 maize. As the first genetically modified seeds were imported from the United States, the debate surrounding Bt maize intensified. Moreover, the mad cow disease crisis became a major issue on both the policy and public agendas. As noted by Toke and others, this crisis was not the principal reason for Europeans rejecting genetically modified food [71]. Nevertheless, the possible spread of BSE shook Europeans' belief in the trustworthiness of the policy institutions responsible for ensuring the public health, deepened their suspicion of genetically modified foods, and influenced policy decision making in many European countries. Advances in and controversies over biotechnology applications did not translate into increased knowledge about genetic engineering. In fact, the knowledge of biotechnology techniques among Europeans remained relatively low and varied by country, as illustrated by Eurobarometer surveys throughout the 1990s.

While the introduction of biotechnology products continued in the US, most members of the Regulatory Committee of the EU and European Parliament objected to the authorization of genetically modified maize. Although the European Commission eventually allowed the import and cultivation of GM maize in 1997, Austria prohibited its import by invoking the safeguard clause of Directive 90/220, which allows member states to restrict products believed to pose a danger to the health and safety of citizens. Despite the intensity associated with these issues and the emergence of a well organized opposition to fight genetically modified products, as illustrated by the anti-genetically modified product movement of NGOs and other interest groups in France, most of the public within the EU had demonstrated low levels of knowledge concerning genetic engineering [158–161]. Based on a nine-item quiz to measure biotechnology knowledge, the Eurobarometer surveys between 1996 and 2002 indicate a slight overall upward trend in knowledge. However, only three countries, Sweden, Denmark, and the Netherlands, consistently passed the quiz by answering even 60 % of the items correctly, or, as in the case of Sweden in 2002, 70 % [162]. While these trends have not changed significantly, a 2005 Eurobarometer poll showed that 80 % of Europeans were familiar with genetically modified food [163].

Regardless of low biotechnology knowledge among the European public, the EU continued its active policy engagement in the regulation of genetically modified food. In response to North American genetically modified soybeans reaching Europe, the EU, with considerable support from the European Council and Parliament, established specific labeling rules and mandated labeling requirements for most genetically modified food under the Novel Foods Regulation (EC) No. 258/97, Council Directive 97/35/EC, and IP/97/1044. Fully introduced by September 1998, the labeling requirements specified in Directive 97/35/EC and IP/97/1044 not only amended Directive 90/220/EEC but also coincided with the disappearance of genetically modified products throughout Europe. Despite having authorized 18 genetically modified products for commercial use since Directive 90/220/EEC, increasing doubts about the safety of food biotechnology applications convinced 12 of the then-15 member states to oppose the authorization of new genetically modified organisms. Faced with this broad-based opposition, the European Commission, rather than challenging strong anti-GM sentiment, agreed to halt the authorization of genetically modified organisms, paving the way for a de facto GM product moratorium that started in 1998 and lasted until 2004 [164, 165].

Pending reform of Directive 90/220/EC, the Council of Environmental Ministers halted any approval of new GM organisms and began to revamp its regulatory system "to better address the challenges of modern biotechnology" [107]. Countries characterized by low levels of uncertainty tolerance, including Greece, France, Italy, Austria, and Germany, either invoked the safeguard clause to ban GM organisms that had already been approved at the EU level or refused approval of new GM products until the development of stricter risk assessment procedures and the implementation of traceability, liability, and labeling rules. Consumers' unions across Europe echoed these sentiments of opposition. Accordingly, the International Consumer's Organization urged "governments ... [to] require full pre-market evaluation and social and safety impact assessment of GM foods" [158].

These events convinced the EU to expand and revamp the regulatory food safety framework. The 1999 *White Paper on Food Safety* proposed the establishment of an independent European food safety agency modeled after the Food and Drug Administration. Aimed at ensuring consumer health protection in the area of food safety and enabling the agency to draw on independent scientific opinions, this proposal became a functional reality in 2002 with the establishment of the European Food Safety Authority and subsequent formation of the Scientific Committee and Scientific Panels a year later [166]. Furthermore, the EU deemed the procedures that govern the deliberate release of GM organisms into the environment under the Directive 90/220/EEC as environmentally unsound and replaced it with Directive 2001/18/EC (the *Deliberate Release Directive*), which reiterated the safeguard clause, reaffirmed the *precautionary* principle, emphasized *preventive* actions, and introduced an *ethical* dimension to assess GM products. As part of the officially sanctioned notification process, this directive required genetically modified food producers to provide a full environmental risk assessment detailing the foreseeable risks of such products to human health and the environment. Member states were authorized to conduct their own investigation and take into account the ethical implications of marketing genetically modified food [167]. Finally, based on guidelines adopted by the European Commission in 2003, 15 of the 27 EU member states have implemented national strategies for the coexistence of genetically modified crops with their organic counterparts [165].

As determined by advanced search engine results on governmental websites for Health Canada and the European Union, there are currently about 600 EU documents dealing with food biotechnology in contrast to about 200 for Canada. While the U.S. Food and Drug Administration does not allow for tailored online searches regarding the regulations of genetically modified food, there are significantly fewer regulations in the United States compared to Europe [168]. As pointed out by Carter and Gruère, "globally, the EU has the most comprehensive regulations on GM food" [169]. The new, more extensive regulatory framework is a continuation of the EU's latest effort to regulate GM food. The *Food and Feed Regulation*, (EC) 1829/2003, clarified a series of previous regulations and directives that directly or indirectly dealt with genetically modified food, including Regulation (EC) No. 258/97 and Directives 82/47/EEC, 2002/53/EC, 2002/55/EC, 68/19/EEC, and 2001/18/EC. Consisting of 49 articles and one annex, the primary objective of Regulation (EC) 1829/2003 is to "provide the basis for ensuring a high level of protection of human life and health, animal health and welfare, environment and consumer interests in relation to genetically modified food" [170]. As a number of countries including Austria, Greece, Germany, Luxembourg, and Germany maintained a ban on certain genetically modified foods, the latest additions to the regulatory framework focus on the traceability of novel foods throughout the production and distribution process [171, 172].

As the regulatory scope has expanded, statements by the European Commission have begun to address the potential benefits of genetically engineered products [173]. However, this more welcoming stance contrasts with the fact that the European Food Safety Authority has approved only one genetically modified product since 2004 [174]. Moreover, public perception of genetically modified food has remained negative. Eurobarometer surveys show that national attitudes toward genetically modified food have been mostly characterized by negative undertones, judging such products as not being useful and a risk to society at large. Large segments of the public in Belgium, Sweden, Denmark, Germany, Italy, the Netherlands, Austria, and Greece were particularly unsupportive of genetically modified food. Except for Denmark, Sweden, and Austria, support for GM food has declined considerably since 1996. EU averages between 1996 and 2005 derived from the Eurobarometer studies confirm the overall decline in and low support for genetically modified food between 1996 and 2005. To some extent, these patterns are also visible in the United States and Canada (see Fig. 2.1). Studies that highlight the Europeans cautious approach to, and the perceived threat associated with,

Fig. 2.1 Public support for GM food in the US, Canada, Brazil, and EU

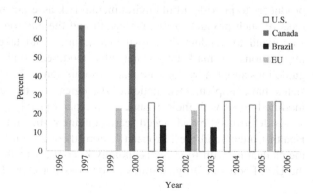

genetically engineered food reiterate the low support for genetically modified food within the EU—a pattern firmly established since the first half of the 1990s [175, 176].

Notes For the U.S., the survey asked: Do you favor the introduction of genetically modified foods into the US food supply? For Canada, the survey asked: Is using biotechnology in the production of food and drinks useful? For Brazil, the survey asked about the preference of genetically modified foods. For the EU, the results for 1996–2002 are based on *decided* Europeans in support of GM food, while the 2005 polling results are based on a combination of those who "agree" and "totally agree" with GM Food [176].

2.5 Conclusion

This study asked whether different uncertainty tolerance levels and risk perceptions provide another explanatory dimension to the formulation of policies regarding genetically engineered food in the United States, Canada, Brazil, and EU. Different uncertainty tolerance levels and risk perceptions among policy stakeholders, defined here as the public and policymakers in relevant regulatory agencies, are linked to distinctive protective policies in the area of food biotechnology. The formulation of stringent regulatory policies occurs within an environment of low uncertainty tolerance levels and prevalent cautious risk perceptions. On the other hand, high uncertainty tolerance levels and the initial prevalence of opportunistic risk perceptions among policy stakeholders encourage the continuation or adjustment of existing protective regulatory policies. The competition of different risk perceptions facilitates the emergence of an inconsistent regulatory framework. While risk perceptions among policy stakeholders can change and remain ambiguous, the findings of this analysis illustrate that country or region-specific differences in political culture—and the prevailing risk perceptions among policy stakeholders associated with them—can add another explanatory dimension to understand policy outcomes (see Table 2.2).

On both sides of the Atlantic, public knowledge about biotechnology applications, including genetically modified food, remained relatively low in the United States, Canada, and the EU throughout the 1990s. However, as skepticism and controversies surrounding genetically modified food deepened and press coverage intensified, familiarity with genetically modified food increased, especially among Europeans and, as indicated by polls conducted in the early 2000s, among Brazilians. Surveys at the state, regional, and international levels also showed that the publics in the United States and Canada were more supportive of food biotechnology applications compared to their European and especially Brazilian counterparts. Because of the perceived benefits and presumed low danger levels of genetically modified foods, a substantial segment of the public in the United States and Canada mostly supported genetically modified food, especially during the first half of the 1990s.

Table 2.2 Genetically modified food policy and policy risk perception timeline

Year	United States	Canada	Brazil	EU
1980	–	MOSST: Biotech-nology in Canada	–	–
1983	–	National Biotech-nology strategy	–	–
1986	*Coordinated framework for regulation of biotechnology*	–	–	–
1987	–	–	–	Viehoff report
1990	–	–	–	Council directive 90/220/EEC on the deliberate release into the environment of GMOs
1992	FDA: *Foods derived from new plant varieties*	–	–	–
1994	FDA approval of the Flavr-Savr™ tomato	HC: *Guidelines for the safety assess-ment of novel foods*	–	–
1995	FDA: *Safety assurance of foods derived by modern biotechnology in the United States*	–	Law of biosecurity establishes the National Biosafety Technical Com-mission (CTNBio)	
1996	–	–	–	European parliament mandates GM food labeling requirements
1997	EPA: Allows lim-ited registration of a new btcorn called star link	Creation of CFIA	–	–
1998	–	Canadian biotech-nology strategy secretariat: *Cana-dian biotechnology strategy: An ongo-ing renewal process*	Federal court prohibits the commercialization of genetically modified soybeans	–
2000	FDA: Confirms traces of Star link in taco shells	–	–	Commission of the European communi-ties: White paper on food safety

(continued)

Table 2.2 (continued)

Year	United States	Canada	Brazil	EU
2001	–	–	–	Council directive 2001/18/EC on the deliberate release of GMOs
2005	–	–	New Law of Biosecurity strengthens the power of the CTN Bio	–
2006	–	–	–	FDA: *Guidance for industry. Recommendations for the early food safety evaluation of new non-pesticidal proteins produced by new plant intended for food use*
Policymaker/Public risk perceptions	1990s			2000s
United States	Opportunistic/opportunistic			Opportunistic/ cautious elements
Canada	Opportunistic/opportunistic			Opportunistic/ cautious elements
Brazil	Competing risk Perceptions/unknown			Competing risk perceptions/cautious
EU	Cautious/cautious			Cautious/cautious

However, by the late 1990s and early 2000s, a public increasingly skeptical and uncertain about such products developed an unfavorable assessment of genetically modified food in the United States, Canada, and the EU, suggesting an opportunistic risk perception mixed with visible signs of caution. While the level of public skepticism changed in the United States, Canada, and the EU, the Brazilian public acknowledged some benefits associated with food biotechnology but was consistently and strongly opposed to it, suggesting a mostly cautious risk perception.

While regulatory adjustments have been proposed in the United States and Canada to reflect the increasing skepticism regarding genetically modified food, policymakers within the Food and Drug Administration and Health Canada have continued to encourage the advancement of genetically engineered food, mostly praising its safety and benefits. In contrast, the consistent suspicion of genetically modified food as something unnatural coincided with the Europeans' less favorable assessment of engineered food products. While exceptions exist, policymakers within the major regulatory bodies of the EU generally downplayed the benefits of genetically modified food and instead emphasized the risks associated with them. Consistent policy trajectories are much more difficult to pinpoint in the case of Brazil. Divisions and competing risk perceptions within government and at the sub-national level dominated the development of Brazil's food biotechnology regulatory framework. Accordingly, the simultaneous and sometimes ambiguous advocacy of risks and benefits in regards to genetically modified food at the federal and state level of government as well as the subsequent legal battles that challenged the Comissão Técnica Nacional de Biossegurança's opportunistic risk perception paved the way for the development of an inconsistent regulatory framework.

Stakeholders in the United States and Canada tended to perceive risks associated with genetically modified food in terms of low threats and high opportunity. Since the late 1990s there has been an increasing and clearly visible cautious risk perception growing among the North American public similar to (although not as severe as) the European and Brazilian outlook. This trend, however, also indicated a widening risk perception gap between the public and policymakers in the United States and Canada. Following a mostly opportunistic risk perception, especially among regulators, within an environment of high tolerance for uncertainty, the United States and Canada adjusted and expanded the responsibilities of the existing protective regulatory frameworks into the area of genetically modified food. Accordingly, the Food and Drug Administration and Health Canada, responsible for the regulation of conventionally produced food, took on the regulatory responsibility for food biotechnology. Within the context of low uncertainty tolerance, the European public and policymakers tended to perceive genetically modified food in terms of high threat and low opportunity. When combined with the fear of the unknown, this mostly cautious risk perception among EU policy stakeholders contributed to the creation of elaborate and stringent protective regulatory policies throughout the EU. Brazil, similar to many European countries characterized by low levels of uncertainty tolerance, initially pursued a precautionary policy approach vis-à-vis food biotechnology. However, as divisions along competing risk perceptions within both the Cardoso and Lula administrations crystallized, partially in response to external and internal pressures, the regulatory framework became increasingly inconsistent.

In addition to conventional explanations that focus on socioeconomic conditions, the role of political institutions, or a process versus product outlook on policy formation, the influence of political culture and risk perceptions provide another useful analytical perspective to understand the genetically modified food policy divide between North America and the EU and, to some extent, the inconsistent policy trajectory of food biotechnology in Brazil. The uncertainty avoidance index and *Margolis Risk Matrix* assists researchers in assessing the influence of differences in political culture and risk perceptions on policymaking. By drawing attention to core values across societies in terms of differences in risk tolerance, these analytical approaches can be reasonably generalized and add to traditional perspectives on policymaking. Nevertheless, conceptual and methodological weaknesses remain regarding the operationalization of the risk perception framework in modeling the dynamic relationship between risk perceptions and other sociopolitical variables. Relying on regulatory policies other than those related to genetically modified food, future studies in this area should refine the political culture/risk perception framework, consider the influence of the media on agenda-setting in the selected policy area, and take into account different policy dynamics as a result of differences in economic and political development.

References

1. Reiss, M., & Straughan, R. (1996). *Improving nature? The science and ethics of genetic engineering.* Cambridge: Cambridge University Press.
2. Agence France Presse. (2012, February 7). *Brazil to lead world in biotech crops.* Sao Paulo: Agence France Presse.
3. Shore, R. (2012, February 11). Biotech agriculture expanding globally: New battle lines are emerging as industry shift from commodity crops to whole foods. *The Vancouver Sun.*
4. Munro, W. A. & Schurman, R. A. (2008). Sustaining outrage: Cultural capital, strategic location, and motivating sensibilities in the US anti-genetic engineering movement. In W. Wright & G. Middendorf (Eds.), *The fight over food. Producers, consumers, and activists challenge the global food system* (pp. 145–176). The Pennsylvania State University: Pennsylvania State University Press.
5. Jasanoff, S. (1995). Product, process, or programme: Three cultures and the regulation of biotechnology. In M. Bauer (Ed.), *Resistance to new technology. Nuclear power, information technology and biotechnology* (pp. 311–331). Cambridge: Cambridge University Press.
6. Dunlop, C. (2000). GMOs and regulatory styles. *Environmental Politics, 9*(2), 149–155.
7. Gaskell, G. (2001). Troubled waters: The Atlantic divide on biotechnology policy. In G. Gaskell & M. Bauer (Eds.), *Biotechnology 1996–2000: The years of controversy* (pp. 96–115). London: NMSI Trading Ltd.
8. Isaac, G. E. (2002). *Agricultural biotechnology and transatlantic trade: Regulatory barriers to GM crops* (p. 211). Cary: CABI Publishing.
9. Katz, P., Macdonald, P., & Mackanzie, G. (2004). The evolving GMO food trade policy debate: Towards a global regulatory regime? In R. E. Evenson & V. Santaniello (Eds.), *The regulation of agricultural biotechnology* (pp. 25–34). Cambridge: CABI Publishing.
10. Jaffe, G. (2004). Regulating transgenic crops: A comparative analysis of different regulatory processes. *Transgenic Research, 13*(1), 5–19.
11. Gaudillière, J. P. (2006). Globalization and regulation in the biotech world: The transatlantic debates over cancer genes and genetically modified crops. *OSIRIS, 21*, 251–272.
12. Sheingate, A. D. (2006). Promotion versus precaution: The evolution of biotechnology policy in the United States. *British Journal of Political Science, 36*(2), 243–268.
13. Jepson, W. E., Brannstrom, C. & de Souza, R. S. (2005). A case of contested ecological modernization: The governance of genetically modified crops in Brazil. *Environment and Planning: Government and Policy, 23*(2), 295–310.
14. Hofstede, G. (1980). *Culture's consequences: International differences in work-related values.* Beverly Hills: Sage.
15. Margolis, H. (1996). *Dealing with risk: Why the public and experts disagree on environmental issues* (pp. 75–78). Chicago: University of Chicago Press.
16. Williams, R. M. (1974). *American Society.* New York: Knopf Williams.
17. Elazar, D. J. (1984). *American federalism: A view from the States.* New York: Harper & Row.

18. Almond, G. A., & Verba, S. (1963). *The civic culture: Political attitudes and democracy in five nations*. Princeton: Princeton University.
19. Deutsch, K. W. (1970). *Politics and government*. Boston: Houghton Mifflin.
20. Swedlow, B. (2002). Toward cultural analysis in policy analysis: Picking up where Aaron Wildavsky left off. *Journal of Comparative Policy Analysis, 4*(3), 267–285.
21. Hoppe, R. (2002). Cultures of public policy problems. *Journal of Comparative Policy Analysis, 4*(3), 305–326.
22. Turner, C. H., & Trompenaars, A. (1993). *The seven cultures of capitalism: Value systems for creating wealth in the United States, Japan, Germany, France, Britain, Sweden, and the Netherlands*. New York: Doubleday Business.
23. Steensma, H. K. (2000). The influence of national culture on the formation of technology alliances by entrepreneurial firms. *The Academy of Management Journal, 43*(5), 951–973.
24. Hofstede, G. (1997). *Cultures and organizations: Software of the mind*. New York: McGraw-Hill.
25. Hofstede, G., Hofstede, G. H., & Hofstede, G. H. (2005). *Culture and organizations: Software of the mind. Intercultural cooperation and its importance for survival* (pp. 167–190). New York: McGraw-Hill.
26. Yniyurt, S., & Townsend, J. (2003). Does culture explain acceptance of new products in a country? An empirical investigation. *International Marketing Review, 20*(4), 377–397.
27. Rowe, W. D. (1992). Risk analysis: A tool for policy decisions. In M. Waterstone (Ed.), *Risk and society: The interaction of science and public policy*. Boston: Kluwer Academic Publishers.
28. Funke, O. (1985). Biopolitics and public policy: Controlling biotechnology. *PS: Political Science and Politics, 18*(1), 69–77.
29. Senge, P. M. (1990). *The fifth discipline: The art and practice of the learning organization*. New York: Currency Doubleday.
30. Kelly, G. A. (1955). *The psychology of personal constructs*. New York: Norton and Sons.
31. Eden, C., & Radford, J. (1990). *Tackling strategic problems*. Newbury Park: Sage.
32. Fransella, F., & Bannister, D. (1977). *A manual for repertory grid technique*. London: Academic Press.
33. Sjöberg, L. (2004). Principles of risk perception applied to gene technology: To overcome the resistance to applications of biotechnology, research on risk perception must take a closer look at the public's reasons for rejecting this technology. *European Molecular Biology Organization, 5*, S47–S51.
34. Slovic, P., Fischoff, B., & Lichtenstein, S. (1976). Facts and fears: Understanding perceived risk. In R. Schwing & W. Albers (Eds.), *Societal risk assessment: How safe is safe enough?* (pp. 181–214). New York: Plenum Press.
35. Nisbett, R., & Ross, L. (1980). *Human inference: Strategies and shortcomings of social judgment*. Englewood Cliffs: Prentice-Hall.
36. Tversky, A., & Kahneman, D. (1986). Rational choice and the framing of decision. *Journal of Business, 59*, 251–278.
37. Slovic, P. (1992). Perception of risk: Reflections on the psychometric paradigm. In S. Krimsky & D. Golding (Eds.), *Social theories of risk* (pp. 117–152). New York: Praeger.
38. Wynne, B. (2001). Creating pubic alienation: Expert cultures of risk and ethics on GMOs. *Science as Culture, 10*(4), 445–481.
39. Anderson, J. E. (1990). *Public policymaking* (p. 5). Boston: Houghton Mifflin.
40. Cobb, R. W., & Elder, C. (1972). *Participation in American politics*. Boston: Allyn & Bacon.
41. Ripley, R. B., & Franklin, G. A. (1986). *Policy implementation and bureaucracy*. Chicago: Dorsey Press.
42. Lipsky, M. (1980). *Street level bureaucracy: Dilemmas of the individual in public services*. Belmont: Russel Sage.
43. The Pew Initiative on Food and Biotechnology. (2006). *Review of public opinion research*. The Mellman Group.

44. Hallman, W. K. et al. (2004). *Americans and GM foods: Knowledge, opinion, and interest in 2004.* New Brunswick: Food Policy Institute, Rutgers, The State University of New Jersey.
45. Hoban, T. J. (2004). *Public attitudes towards agricultural biotechnology, ESA working paper no. 04–09.* Agricultural and Development Economics Division: The Food and Agricultural Organization.
46. Hoban, T. J. (2002). *American consumers' awareness and acceptance of biotechnology.* Washington, DC: National Agricultural Biotechnology Council.
47. Hoban, T. J. (1999). Consumer acceptance of biotechnology in the United States and Japan. *Food Technology, 53,* 85–88.
48. Hoban, T. J. (1999). Public perceptions and understanding of agricultural biotechnology. *Economic Perspectives: An Electronic Journal of the Department of State 4.* Retrieved October 14, 2006, from http://usinfo.state.gov/journals/ites/1099/ijee/bio-hoban.htm.
49. Hoban, T. J. (1997). Consumer acceptance of biotechnology: An international perspective. *Nature Biotechnology, 15,* 232–234.
50. Hoban, T. J. (1996). Trends in consumer acceptance and awareness of biotechnology. *Journal of Food Distribution Research, 27*(1), 1–10.
51. Hoban, T. J. (1990). *Agricultural biotechnology: Public attitudes and educational needs.* Raleigh: North Carolina State University.
52. Schilling, B. J. et al. (2002). *Consumer knowledge of food biotechnology. A descriptive study of US residents.* Rutgers University: Food Policy Institute.
53. Wolf, M. & Domegan, C. (2002). A comparison of consumer attitudes towards GM food in Ireland and the United States: A case study over time. In V. Santaniollo, R. E. Evenson, & D. Zilberman (Eds.), *Market development for genetically modified foods.* (pp. 25–38). New York: CABI Publishing.
54. Hallman, W. K. & Metcalfe, J. (1994). *Public perceptions of agricultural biotechnology: A survey of New Jersey residents.* Food Policy Institute, Cook College, Rutgers, The State University of New Jersey.
55. Hoban, T. J., & Kendall, P. (1992). *Consumer attitudes about the use of biotechnology in agriculture and food production.* Raleigh, North Carolina: North Carolina State University.
56. Decima Research. (2006). *Emerging technologies tracking.* Canada: Decima Research.
57. Research, Pollara. (2003). *Public opinion research into biotechnology issues in Canada.* Toronto: Pollara Research.
58. Pollara Research & Earnscliffe Research and Communications. (2000). *Public opinion research into biotechnology issues.* Ottawa: Earnschliffe Research and Communications Pollara Research and Earnscliffe.
59. Einsiedel, E. F. & Medlock, J. M. (2001). Canada on the gene trail. In G. Gaskell & M. W. Bauer (Eds.), *Biotechnology 1996–2000. The years of controversy* (pp. 145–156). London: NMSI Trading Ltd.
60. IBOPE. (2001). *Pesquisa de opiniao publica sobre transgenicos.* Retrieved March 14, 2012 from http://www.greenpeace.com.br/transgenicos/pdf/pesquisaIBOPE_agosto2001.pdf.
61. IBOPE. (2002). *Pesquisa de opiniao publica sobre transgenicos.* Retrieved March 15, 2012 from http://www.ibope.com.br/opp/pesquisa/opiniaopublica/download/opp573_transgenicos.pdf.
62. IBOPE. (2003). *Pesquisa de opiniao publica sobre transgenicos.* Retrieved March 15, 2012, from http://www.greenpeace.org.br/transgenicos/pdf/pesquisaIBOPE_2003.pdf.
63. Massarani, L., & de Castro Moreira, I. (2005). Attitudes towards genetics: A case study among Brazilian high school students. *Public Understanding of Science, 14*(2), 201–212.
64. Massarani, L., & de Castro Moreira, I. (2006). What do Brazilians think about transgenics. In D. Brossard, J. Shanahan, & T. C. Nesbitt (Eds.), *The public, the media, and agricultural biotechnology* (pp. 179–1990). Cambridge: CABI.
65. Oosterveeer, P. (2007). *Global governance of food production and consumption. Issues and challenges* (p. 120). Cheltenham: Edward Elgar Publishing Limited.

66. Food and Drug Administration. (1992). Statement of policy: Foods derived from new plant varieties. *Federal Register, 57*, 22984–23005.

67. Lacy, W. B., Busch, L. & Lacy, L. L. (1991). Public perceptions of agricultural biotechnology. In B. R. Baumgardt & M. A. Martin (Eds.), *Agricultural biotechnology. Issues and choices.* (pp. 139–161). West Lafayette: Purdue University Agricultural Experiment Station.

68. Hoban, T. J. (1998). International acceptance of agricultural biotechnology. In R. W. F. Hardy & J. B. Segelken (Eds.), *Agricultural biotechnology and environmental quality: Gene escape and pest resistance* (p. 61). Ithaca: National Agricultural Biotechnology Council Report.

69. Baram, M. (2010). Governance of GM crops and food safety in the United States. In M. Baram & M. Bourrier (Eds.), *Governing risk in GM agriculture* (pp. 16–24). Cambridge: Cambridge University Press.

70. Ferrara, J. (2001). Paving the way for biotechnology: Federal regulations and industry PR. In B. Tokar (Ed.), *Redesigning life? The worldwide challenges to genetic engineering* (pp. 297–305). London: Zed Book Ltd.

71. Toke, D. (2004). *The politics of GM food: A comparative study of the UK, USA and EU* (pp. 107–157). London: Routledge.

72. Barbee, M. (2004). *Politically incorrect nutrition: Finding reality in the mire of food industry propaganda* (p. 70). Ridgefield: Vital Health Publishing.

73. Harrington, J. C. (2005). *The challenge to power. Money, investing, and democracy.* White River Junction: Chelsea Green Publishing Company.

74. Pelaez, V. & Schmidt, W. (2004). Social struggles and the regulation of transgenic crops in Brazil. In K. Jansen & S. Vellema (Eds.), *Agribusiness and society. Corporate responses to environmentalism, market opportunities and public regulation.* (pp. 232–246). London: Zed Books.

75. Reilly, J. M. (1989, December). Consumer effects of biotechnology. *Agriculture Information Bulletin, 581*, 1–11.

76. West, D. M. (2007). *Biotechnology policy across national boundaries. The science-industrial complex* (pp. 51–60). New York: Palgrave MacMilland.

77. Food and Drug Administration (1993, January/February). Genetically engineered foods: Fears & facts. An interview with FDA's Jim Maryanski, *FDA Consumer.*

78. Food and Drug Administration (2010). Completed consultations on bioengineered foods. Retrieved May 1, 2010, from http://www.accessdata.fda.gov/scripts/fcn/fcnNavigation.cfm?rpt=bioListing.

79. Vogt, D. V. (2010). *Food biotechnology in the United States: Science, regulation, and issues* (p. CRS 9). Washington, DC: CRS Report for Congress, Domestic Social Policy Division. Retrieved May 10, 2010, from http://www.whprp.org/NLE/CRSreports/science/st-41.pdf.

80. Food and Drug Administration (1994, May 18). *Biotechnology of food*, FDA backgrounder BG94-4. Washington DC: Center for Food Safety and Applied Nutrition.

81. Food and Drug Administration (1994, September). First biotech tomato marketed. *FDA Consumer Magazine*, 3–4.

82. Hoban, T. J. (1999). Consumer acceptance of biotechnology in the United States and Japan. *Food Technology, 53*(5), 50–53.

83. The Pew Initiative on Food and Biotechnology. (2006). Genetically modified crops in the United States.. p. 3.

84. Zimmerman, L., et al. (1994). Consumer knowledge and concern about biotechnology and food safety. *Food Technology, 48*(11), 71–77.

85. Shanahan, J., Scheufele, D., & Lee, E. (2001). The polls—trends: Attitudes about agricultural biotechnology and genetically modified organisms. *Public Opinion Quarterly, 65*, 267–281.

86. Winston, M. (2002). *Travels in the genetically modified zone* (p. 236). USA: President and Fellows of Harvard College.

87. Gaskell, G., et al. (2000). Biotechnology and the European public. *Nature Biotechnology, 18* (September), 935–938.

88. Food and Drug Administration. (1996). *Safety assurance of foods derived by modern biotechnology in the United States.* Washington, DC: Center for Food Safety and Applied Nutrition.
89. Food and Drug Administration. (1995). *FDA's Policy for foods developed by biotechnology.* Washington, DC: Center for Food Safety and Applied Nutrition.
90. Bucchini, L., & Goldman, L. R. (2002). Starlink corn: A risk analysis. *Environmental Health Perspectives, 110*(1), 5–13.
91. Food and Drug Administration. (1999, October 7). *Statement on biotechnology issues by James H. Maryanski, Ph.D. Biotechnology Coordinator, Center for Food Safety and Applied Nutrition before the Senate Committee on Agriculture, Nutrition and Forestry.* Retrieved September 15, 2006, from http://agriculture.senate.gov/Hearings/Hearings_1999/mar99107.htm.
92. Food and Drug Administration. (2000, September 26). *Statement by Joseph A. Levitt, Esq. Director, Center for Food Safety and Applied Nutrition. Food and Drug Administration Department of Health and Human Services before the Health, Education, Labor, and Pensions Committee.* Retrieved September 15, 2006, from http://www.biotech-info.net/JAL.html.
93. Food and Drug Administration. (2000, January/February). Are bioengineered foods safe? *FDA Consumer Magazine.* Retrieved September 25, 2006, from http://www.agbioworld.org/biotech-info/articles/biotech-art/fda_mag.html.
94. Department of Health and Human Services. (2001, January 18). Premarket notice concerning bioengineering foods (proposed rule). 66 Federal Register 12, pp. 4706–4738. Retrieved November 12, 2009, from http://www.fda.gov/Food/GuidanceComplianceRegulatory Information/GuidanceDocuments/Biotechnology/ucm096149.htm.
95. Food and Drug Administration. (2014). Biotechnology consultations on food from GE plant varieties. Retrieved December 9, 2014, from http://www.accessdata.fda.gov/scripts/fdcc/?set=Biocon.
96. Food and Drug Administration. (2004). *Talk paper. FDA proposes draft guidance for industry for new plant varieties intended for food use.* Washington, DC: Center for Food Safety and Applied Nutrition.
97. Food and Drug Administration. (2006). *Guidance for industry. Recommendations for the early food safety evaluation of new non-pesticidal proteins produced by new plant varieties intended for food use.* Washington, DC: Food and Drug Administration Center for Food Safety and Applied Nutrition.
98. Carter, C. A., & Gruère, G. P. (2006). International approval and labeling regulations of genetically modified food in major trading countries, in regulating agricultural Biotechnology. In R. E. Just, J. M. Alston, & D. Zilberman (Eds.), *Economic and policy* (p. 467). New York: Springer Science+Business Media.
99. Food and Drug Administration. (2001). *Guidance for industry, voluntary labeling indicating whether foods have or have not been developed using bioengineering.* Washington DC: Center for Food and Applied Nutrition.
100. Food and Drug Administration and US Department of Agriculture. (2002). *Guidance for industry: Drugs biologics and medical devices derived from bioengineered plants for use in humans and animals—draft guidance.* Washington, DC: US Food and Drug Administration and United States Department of Agriculture.
101. Bren, L. (2000, November-December). Genetic engineering: The future of foodss? *FDA Consumner Magazine.*
102. Murphy, D. J. (2007). Improving containment strategies in biopharming. *Plant Biotechnology Journal, 5,* 555–569.
103. Abergel, E., & Barrett, K. (2002). Putting the cart before the horse: A review of biotechnology policy in Canada. *Journal of Canadian Studies, 37*(3), 137–141.
104. Standing Committee on Agriculture and Agri-Food. (2003). *Hearing. June 12, 2003.* Standing Committee on Agriculture and Agri-Food.

105. Andree, P. (2002). The biopolitics of genetically modified organisms in Canada. *Journal of Canadian Studies, 37*(3), 162–191.
106. Standing Committee on Agriculture and Agri-Food. (1998, May 12). *Evidence*. Standing Committee on Agriculture and Agri-Food.
107. Andree, P. (2006). GM food regulation. An analysis of efforts to improve genetically modified food regulation in Canada. *Science and Public Policy, 33*(5), 377–389.
108. Skogstad, G. (2010). Science and technology: Politicians and the public. In J. C. Courtney & D. E. Smith (Eds.), *The Oxford handbook of canadian politics* (p. 460). Oxford: Oxford University Press.
109. Canadian Biotechnology Advisory Committee. (2002). Improving the regulation of genetically modified foods and other novel foods in Canada. Ottawa: Canadian Biotechnology Advisory Committee. Retrieved October 21, 2009, from http://dsp-psd. pwgsc.gc.ca/Collection/C2-589-2001-1E.pdf.
110. Grant, E. I., & Hobbs, J. E. (2002). GM food regulations: Canadian debates. *Canadian Journal of Policy Research, 3*(2), 105–113.
111. Health Canada. (1994). *Guidelines for the safety assessment of novel foods. Preamble and guidance scheme for notification* (Vol. 1). Ottawa: Food Directorate Health Protection Branch Health Canada. Retrieved November 1, 2009, from http://www.hc-sc.gc.ca/fn-an/alt_ formats/hpfb-dgpsa/pdf/legislation/nvvli-eng.pdf.
112. Health Canada. (2010). *Approved products*. Retrieved May 5, 2010, from http://www.hc-sc. gc.ca/fn-an/gmf-agm/appro/index-eng.php.
113. Health Canada. *Information: safety assessment of the Flavr Savr tomato*. Retrieved September 14, 2006, from http://www.hc-sc.ca/fn-an/gmf-agm/appro/favr_savtm_tomato_ tomate_flavr_savrmd_e.html.
114. Pollara Research. (2003). Public opinion research into biotechnology issues in Canada. Toronto: Pollara Research. p. 4.
115. Health Candada. (2006). *Guidelines for the safety assessment of novel foods*. Retrieved December 9, 2014, from http://www.hc-sc.gc.ca/fn-an/alt_formats/hpfb-dgpsa/pdf/gmf-agm/ guidelines-lignesdirectrices-eng.pdf.
116. Canadian Biotechnology Strategy Secretariat. (1998). *Canadian biotechnology strategy: An ongoing renewal process*. Ottawa: Canadian Biotechnology Strategy Secretariat. Retrieved April 4, 2010, from http://www.bioportal.gc.ca/ENGLISH/View.asp?pmiid=520&x=535.
117. Department of Foreign Affairs and International Trade. (2003). *Biotechnology transforming society: Creating an innovative economy and a higher quality of life. Report on biotechnology 1998–2003*. Government of Canada.
118. Codex Alimentarius Commission. (2000). *Report of the first session of the codex ad hoc intergovernmental task force on foods derived from biotechnology*. Rome: UN/FAO Headquarters. Retrieved May 1, 2010, from ftp://fao.org/codex/ALINORM01/Al01_34e.pdf.
119. Skogstad, G. (2008). *Internationalization and Canadian agriculture. Policy and governing paradigms* (p. 230). Toronto: Toronto University Press.
120. de Castro, L. R. (2011). *Big issues around a tiny insect: Discussing the release of genetically modified mosquitoes (GMM) in Brazil and beyond*. MA European Studies of Society, Science, and Technology. Retrieved April 2, 2012, from http://esst.eu/wp-content/uploads/ LousiaCastroMasterThesis.pdf.
121. Brasil. (1988). Constituição da República Federativa do Brasil. Brasília, D.F.
122. Rhodes, R. (2007). *The politics of agricultural biotechnology in Brazil, 1995–2005. Interests and institutions as explanations for unexpected policy choices*. Paper presented at the 48th Annual Convention of the International Studies Association. Chicago, pp. 19–20.
123. USDA Foreign Agricultural Service. (2010). Brazil: Biotechnology–GE plants and animals. Brazilian annual biotechnology production and putlook. *Global Agricultural Information Network*. Retrieved April 2, 2012, from http://gain.fas.usda.gov/Recent%20GAIN% 20Publications/Biotechnology%20%20GE%20Plants%20and%20Animals_Brasilia_Brazil_ 7-23-2010.pdf.

124. Jepson, W. E. (2002). Globalization and Brazilian biosafety: The politics of scale over biotechnology governance. *Political Geography, 21*, 912.
125. CTNBio report cited in Pelaez, V. & Schmidt, W. Social struggles and the regulation of transgenic crops in Brazil. (p. 247).
126. Pelaez, V. (2009). State of exception in the regulation of genetically modified organisms in Brazil. *Science and Public Policy, 36*(1), 61–71.
127. Branford, S. (2001, September 5). Society: Environment: Bean stalked: Brazil is the only major world producer of GM-free soya, but for how long? *The Guardian*, 8.
128. McAllister, L. K. (2005). *Judging GMOs. Judicial application of the precautionary principle in Brazil*. The National Agricultural Law Center. Retrieved April 8, 2012, from http://www.nationalaglawcenter.org/assets/bibarticles/mcallister_judging.pdf.
129. Rhodes, S. (2006). *The politics of agricultural biotechnology in Brazil, 1995–2005. Interests and institutions as explanations for unexpected policy choices*. Presented at the Congress of the Latin American Studies Association, San Juan, Puerto Rico, p. 2.
130. BBC. (2000). Court ruling on GM products orders environmental impact study. *BBC Summary of World Broadcasts*, July 8, S1.
131. Jepson, W. Brannstrom, C. E. & de Souza, R. S. (2008). Brazilian biotechnology governance: Consensus and conflict over genetically modified crops. In G. Otero (Ed.), *Food for the few. Neoliberal globalism and biotechnology in Latin America*. (pp. 221–305). Austin: University of Texas Press.
132. Prudente, P. S. cited in Bauer, M. W. (2006). The paradoxes of resistance. In G. Gaskell & M. W. Bauer (Eds.), *Genomics and society: Legal, ethical, and social dimensions*. (p. 231). London: Earthscan.
133. BBC. (2000, July 15). Ministers defend genetically modified products. *BBC Summary of World News*, S1.
134. Government communiqué cited in Pelaez V. & Schmidt W. (2004). p. 249.
135. Griesse, M. A. (2007). Developing social responsibility: Biotechnology and the case of DuPont in Brazil. *Journal of Business Ethics, 73*, 111.
136. Bauer, M. W. (2006). The paradoxes of resistance. In G. Gaskell & M. W. Bauer (Eds.), *Genomics and society: Legal, ethical, and social dimensions* (pp. 1–20). London: Earthscan.
137. Pelaez, V., & da Silva, L. R. (2009). Issues concerning a sustainable regulatory body for GMO's in Brazil. *Journal of Economic Issues, XLIII*, 2, 489.
138. Smith, T. (2003, October 15). Brazil's farmers cheer legalization of their gene-altered soy crops. *The International Herald Tribune*, 16.
139. Brazil grants GM soy temporary approval. (2003, April 2–9). *Chemical Week*, 51.
140. Sur, N. (2003). Brazil government lifts ban on transgenic soy. *NACLA Report on the Americas, XXXVII*(3), 2, 5.
141. Leite, M. (2004). *Communicating biotechnology in Brazil: The failure of scientific and public proofs among widespread anti- and pro-technoscientific fundamentalisms*. Paper presented at the 4S-EASST Conference, Paris.
142. Farias, P., Leite, J., & Allain, J. M. (2011). Evolution of the regulatory system for GM crops in Brazil. In M. Baram & M. Bourrier (Eds.), *Governing risk in GM agriculture* (pp. 113–136). Cambridge: Cambridge University Press.
143. Smith, T. (2003, September 26). Brazil lifts ban on gene-altered soy crops. *The International Herald Tribune*, 12.
144. Silva, J. F. (2011). Brazil. Agricultural biotechnology annual. Biotech annual 2011. *Global Agricultural Information Network*, BR 0713.
145. OECD. (1986). *Recombinant DNA safety considerations. Safety considerations for industrial, agricultural and environmental applications of organisms derived by recombinant DNA techniques*. Paris: OECD. Retrieved May 4, 2010, from http://www.oecd.org/dataoecd/43/34/40986855.pdf.
146. Levidow, L. (2001). Precautionary uncertainty: Regulating GM crops in Europe. *Social Studies of Science, 31*(6), 842–874.

147. Löfstedt, R. E., Fischhoff, B., & Fischhoff, I. (2002). Precautionary principles: General definitions and specific applications to genetically modified organisms. *Journal of Policy Analysis and Management, 21*(3), 381–407.
148. Christoforou, T. (2004). The precautionary principle, risk assessment, and the comparative role of science in the European Community and the US legal system. In N. J. Vig & M. G. Faure (Eds.), *Green giants? Environmental policies of the United States and the European Union.* (pp. 17–52). Cambridge: The MIT Press.
149. European Communities. (1990). Council Directive 90/219/EEC of 23 April 1990 on the contained use of genetically modified micro-Organisms. *Official Journal of the European Communities,* L 117.
150. European Communities. (1990). Council Directive 90/220/EEC of 23 April 1990 on the deliberate release into the environment of genetically modified organisms. *Official Journal of the European Communities,* L 0015–0027.
151. European Commission. (2006). *EU Policy on biotechnology.* Luxembourg: European commission environment DG. Retrieved March 24, 2010, from http://ec.europa.eu/environment/biotechnology/pdf/eu_policy_biotechnology.pdf.
152. Pollack, M. A., & Shaffer, G. C. (2000). Biotechnology: The next transatlantic trade war. *Washington Quarterly, 23*(4), 41–54.
153. Anderson, D. R. (2000). Biotechnology risk management: The case of genetically modified organisms (GMOs). *CPCU Journal, 54,* 215–230.
154. Pew Initiative on Food and Biotechnology (2003). *US versus EU: An examination of the trade issues surrounding genetically modified food.* Pew Initiative on Food and Biotechnology Pew Initiative on Food and Biotechnology.
155. Byrne, D. (2002). *Speech 02/314. David Byrne. European commissioner for health and consumer protection. GM food and feed.* Strasbourg: Plenary Session of the European Parliament.
156. Byrne, D. (2001). *Speech 01/378. David Byrne. European commissioner for health and consumer protection. Proposal for a regulation on GM food and feed.* Brussels: European Parliament.
157. Marris, C., et al. (2001). *Public perceptions of agricultural biotechnology in Europe. Final report of the PABE research project.* Brussels: Commission of European Communities.
158. Bonny, S. (2003). Why are most Europeans opposed to GMOs? Factors explaining rejection in France and Europe. *Electronic Journal of Biotechnology, 6*(1), 52–71.
159. Toffel, M. W., & Heyman, J. (2002). *An Atlantic divide? European and American attitudes on genetically engineered food.* University of California, Haas School of Business: California.
160. Verdurme, A., et al. (2001). Differences in public acceptance between generic and premium branded GM food products: An analytical model. In V. Santaniollo, R. E. Evenson, & D. Zilberman (Eds.), *Market development for genetically modified foods* (pp. 39–47). New York: CABI Publishing.
161. Hanf, C. H. & Böker, A. (2001). Is European consumers' refusal of GM food a serious obstacle or a transient fashion? In V. Santaniollo, R. E. Evenson, & D. Zilberman (Eds.), *Market development for genetically modified foods.* (pp. 49–54). New York: CABI Publishing.
162. Gaskell, G., Allum, N., & Stares, S. (2002). *Europeans and biotechnology in 2002: Eurobarometer* (p. 22). Retrieved September 3, 2009, from http://ec.europa.eu/public_opinion/archives/ebs/ebs_177_en.pdf.
163. Gaskell, G. et al. (2006). *Europeans and biotechnology in 2005: Patterns and trends* (p. 16, 60). Retrieved September 3, 2009, from http://ec.europa.eu/research/press/2006/pdf/pr1906_eb_64_3_final_report-may2006_en.pdf.
164. Lee, M. (2008). *EU regulation of GMOs. Law and decision making for a new technology* (pp. 2–4). Chelensham: Edward Elgar Publishing.

165. Schauzu, M. (2010). The European Union's regulatory framework. Developments in legislation, safety assessment, and public perception. In M. Baram & M. Bourrier (Eds.), *Governing risk in GM agriculture* (pp. 57–84). Cambridge: Cambridge University Press.
166. Commission of the European Communities. (2000). *White paper on food safety.* Brussels: Commission of the European Communities. Retrieved March 2, 2010, from http://ec.europa.eu/dgs/health_consumer/library/pub/pub06_en.pdf.
167. Commission of the European Communities. (2001). *Communication from the Commission. Towards a strategic vision of life sciences and biotechnology. Consultation document.* Brussels: Commission of the European Communities. Retrieved March 24, 2010, from http://ec.europa.eu/biotechnology/docs/doc_en.pdf.
168. Stemke, D. J. (2004). Genetically modified organisms: Biosafety and ethical issues. In S. R. Parekh (Ed.), *The GMO handbook. genetically modified animals, microbes and plants in biotechnology* (p. 108). Totowa: Humana Press.
169. Carter, C. A. & Gruère, G. P. (2006). International approval and labeling regulations of genetically modified food in major trading countries. In R. E. Just, J. M. Alston, & D. Zilberman (Eds.), *Regulating agricultural biotechnology. Economic and policy* (p. 470). New York: Springer Science+Business Media.
170. European Union. (2003). Regulation (EC) No 182'9/2003 of the European Parliament and of the Council of 22 September 2003 on genetically modified food and feed. *Official Journal of the European Union.* Department of Health and Human Services, L268/5.
171. Shaffer, G. C., & Pollack, M. A. (2009). The EU regulatory system of GMOs. In M. Everson & E. Vo (Eds.), *Uncertain risks regulated* (p. 280). New York: Routledge-Cavendish.
172. Tzotzos, G. T., Head, G. P., & Hull, R. (2009). *Genetically modified plants. Assessing safety and managing risk* (pp. 138–143). San Diego: Academic Press.
173. European Commission. (2004). *Report from the commission to the European parliament, the council and the European economic and social committee. Life sciences and biotechnology— A strategy for Europe second progress report and future orientations.* Brussels: Commission of the European Communities. Retrieved May 20, 2010, from http://ec.europa.eu/biotechnology/docs/com2004-250_en.pdf.
174. European Commission. (2006). *List of the genetically modified material which has benefited from a favorable risk evaluation within the meaning of Article 47 of regulation (EC) No 1829/2003.* Luxembourg: European Commission. Retrieved May 20, 2010, from http://ec.europa.eu/food/food/biotechnology/gmfood/events_en.pdf.
175. Anderson, L. (2000). *Genetic engineering, food, and our environment.* White River Junction: Chelsea Green Publishing.
176. Peters, H., et al. (2007). Culture and technological innovation: Impact of institutional trust and appreciation of nature on attitudes towards food biotechnology in the USA and Germany. *International Journal of Public Opinion Research, 19*(2), 191–220.

Chapter 3
The Importance of Risk Communication as an Integral Part of Risk Management in the Republic of Serbia

Vesela Radovic and Jean-Marc Mercantini

Abstract Risk management is heavily dependent on information quality (e.g. reliability, pertinence) for making timely and efficient decisions. Obviously, information deficiencies will negatively affect the whole risk management organization and may have also negative impacts on the population, which rapidly may evolve into uncontrollable behaviours. Risk communication is usually considered to be one of the important phases within the risk management process. To illustrate this importance, the case of a country (the Republic of Serbia) where the risk communication system shows many deficiencies, is presented and analysed. One of the results is that message processing depends on the cognitive characteristics and cognitive limitations of the recipients, as well as message properties. It is of prime importance that policy makers keep these two considerations in mind. The most critical result observed in this case study is when citizens are not any more in trust with their authorities. The Republic of Serbia is facing great challenges to design its new risk communication system. The chapter engages a set of questions for public discussions. It initiates the government and other important actors to manage risks and to communicate about them more efficiently, keeping in mind that these actions affect local, regional, national and international relations. By the use of cognitive approaches, it is suggested to the stakeholders to design new organizations and new methodological tools, which could help them to solve problems in the domain of the risk and increase the overall security in the Serbian society.

V. Radovic (✉)
Faculty of Applied Security, Educons University, Vojvode Putnika 87,
21 208 Sremska Kamenica, Serbia
e-mail: veselaradovic@yahoo.com

J.-M. Mercantini
Laboratoire des Sciences de l'Information et des Systèmes,
Aix-Marseille Université, Avenue Escadrille Normandie-Niemen,
13397 Marseille cedex 20, France
e-mail: jean-marc.mercantini@Lsis.org

© Springer-Verlag Berlin Heidelberg 2015   61
J.-M. Mercantini and C. Faucher (eds.), *Risk and Cognition*,
Intelligent Systems Reference Library 80, DOI 10.1007/978-3-662-45704-7_3

3.1 Introduction

Risk management is a rapidly developing discipline in the world of science. Therefore, researchers interested in this matter could rapidly propose various views and descriptions of what risk management involves, how it should be conducted and what it is for. Many researchers have implied that risk is not fact, but a composite of values, specific contexts, and future events [1]. Being at risk is the way of being and ruling in the world of modernity; being at global risk is the human condition at the beginning of the twenty-first century. The more emphatically the existence of world risk society is denied, the more easily it can become a reality [2]. It is necessary to have this fact in mind to find a way to respond adequately on these global risks. The different risk management standards implemented in the world have been created as tools within a process of achieving adequate response to the risks. One of the main constraints of the risk management process is to obtain pertinent information for making a correct, timely and realistic decision for a well being of the society. And, as notified by researchers, especially in social sciences, risk communication is one of the important phases within the risk management process. The field of risk communication has been developing since the late eighties. In terms of their history and organization, the fields of communication and cognitive science share many characteristics. By building theoretically driven, empirically tested structures of cognitive processes, cognitive scientists seek to increase understanding of the mind, as well as to build systems that are able to understand, predict, and generate human thought and action (Information Processing) [3]. Scientists try to answer the questions: how information processes are represented in the mind, and how people react during an emergency. Plough and Krimsky pointed in 1987 'the emergence of risk communication as significant new organizing theme for a set of diverse, but conceptually related problems concerning the political management of public risk perceptions and individual behavioural responses to risk [4].

There are numerous definitions of risk communication, as well as many new scientific concepts regarding different circumstances in which it has to be applied. For the purpose of this chapter authors have chosen the definition from the Joint Project of the World Health Organization (WHO) and the Organisation for Economic Cooperation and Development (OECD). This document defines risk communication as the interactive exchange of information about (health or environmental) risks among risk assessors, managers, news media, interested groups and the general public [5]. Peter Sandman, a noted risk communication expert who has advised companies and governments about various communication crises stated that "the current version of risk communication was born to guide the new partnership and dialogue of government and industry with public" [6]. Risk communication is a subject tackles by some social science theories and models are proposed to explain how people think, reason, make choices and within in emergency situations. Some of these theories are about social learning, group decision-making, decision analysis, etc.

Research from the risk communication field draws heavily on social, cognitive and economic psychology, and their organisational and community-based applications

[7]. Risk communication field like cognitive science, has its roots in numerous disciplines such as psychology, sociology, political science, linguistics, journalism, anthropology, and economics. Therefore, research departments in cognitive science and communication are becoming more common. Research works in decision theory could be used to predict people reactions in emergency situations, for example with the specific perspective to study the influence of the mental representation of some threats on risk communication. Numerous factors have contributed to the rapid growth of the risk communication field, and the contemporary world under the governance of information and communication technologies is the arena for risk communication among interested stakeholders.

According to [5], the framework of the risk analysis process consists of three components: risk assessment, risk management, and risk communication (Fig. 3.1). It involves a logical and systematic method of defining the concepts of risk assessment, risk management and risk communication, where activities, functions or processes are implemented with the purpose to minimize the level of the risk (environmental, health, etc.).

As defined in the IPCS Harmonization Project [5], but also in the International Norm on Risk Management (ISO/FDIS 31000), risk communication appears to be one of the fundamental processes for making correct, timely and realistic decisions for the well being of the society. Risk communication has to be viewed as a permanent, continual and iterative process intended to provide, share or obtain information. As previously said, this information is the base of the decision making process but it must be also the base for new educational programs.

Serbia is a country exposed to various risks [8]. The current circumstances and past inheritances have conducted policy makers to establish an integral risk management policy. The significant issue now is how to create prerequisites for its implementation but also how to provide the required number of scholars in cognitive science who could teach and develop researches about communication.

The purpose of this chapter is to illustrate the challenges facing Serbia to design its risk communication system. It presents the current state of the risk management and of the risk communication with the focus on failures. It engages a set of questions for public discussion, it initiates the government and other important actors to manage risks and to communicate about them more efficiently, keeping in

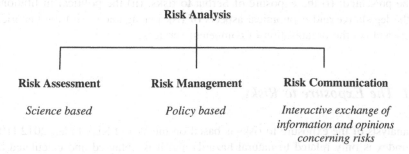

Fig. 3.1 Risk analysis framework

mind that these actions affect local, regional, national and international relations. By the use of cognitive approaches, it is suggested to the stakeholders to design new systems and new methodological tools, which could help them to solve problems in the domain of the risk and increase the overall security in the Serbian society.

The methodology used in the chapter to study the importance of risk communication, as an integral part of risk management is coherent with social science approaches: historical analysis, comparative analysis, and document analysis. The authors have reviewed literature from various disciplines that inform readers about issues of the research: cognitive science, educational science, risk management, risk communication, disaster management, etc. A corpus of documents has been constituted. It is made of electronic databases, books, scientific journals, official documents and examples of positive practices from international communities, as well as numerous syllabuses from the USA universities, publications of the most influential international organizations in the area of risk management and risk communication.

The organisation of the chapter is divided into five parts. In the first part (paragraph number two), the current state of the risk management in Serbia is presented. In the second part (paragraph number three), a discussion about the importance of the risk communication at the local level is developed. The third part (paragraph number four) is devoted to a case study (implementation of a hazardous waste policy) to illustrate the lacks and the failures of the risk communication at the local level in Serbia. In the fourth part (paragraph number five), a set of opportunities are proposed and discussed for the future improvements of the risk communication. The conclusion is the fifth part of the chapter.

3.2 The Risk Management in Serbia

For a part, the modern states' legitimacy derives from their ability to prevent and protect the population from potential physical harms that could emerge from risky situations. During these situations, risk management and risk communication in Serbia were evaluated as inadequate by some of the most influent institutions [9]. Each year, the Government provides large amounts of aid to the citizens affected by various risks. In this part of the chapter the following aspects of the current situation will be presented: (i) the exposure of Serbia to risks, (ii) the political institutions, (iii) the legislative and economical aspects of risk management, (iv) the historical aspects and (v) the organisation of competent services.

3.2.1 The Exposure to Risks

The analysis of the exposure to risks is based on the World Risk Index 2012 [10]. This index is only related to natural hazards and it is obtained and calculated by combining the four components of exposure, susceptibility, lack of coping capacities

Table 3.1 The World Risk Index 2012 for selected countries

Country	Rank	Vulnerability (%)	Exposure (%)	World Risk Index (%)
Albania	38	46.89	21.25	9.96
Serbia	66	42.52	18.05	7.67
Greece	72	34.83	21.11	7.35
Romania	82	42.99	15.77	6.78
Bosnia and Herzegovina	86	47.31	14.02	6.63
FRY Macedonia	95	43.47	14.38	6.25
Hungary	102	37.61	15.61	5.87
Bulgaria	118	39.11	11.66	4.56
Croatia	123	37.73	11.53	4.35
Slovenia	132	32.86	11.59	3.81

The value is the result of the product of Exposure and Vulnerability (the two contributed factors)
Source World Risk Report 2012

and lack of adaptive capacities. The last three components describe the societal aspects of the risk, and combined, they yield the vulnerability factor. It is legitimate to suppose that the situation would be worst if the study could be extended to anthropogenic hazards. The Table 3.1 shows that Serbia has a high possibility of being affected by natural disasters. It holds the second place in the selected set of the neighbour countries, with high values for both factors of exposure and vulnerability. Serbia is at the 66th position out of 173 countries that are included in the World Risk Index overview [10].

This bad evaluation of the country organization facing risks is due in part to the bad economical situation, which can be explained by the recent events that have affected the country: (i) the period of the national conflicts, (ii) the international sanctions, (iii) the economical deterioration caused by the privatisation process and (iv) the global economical crisis. The bad economical situation of the country is characterized by: (i) the rising prices, (ii) the increase of the inflation rate, (iii) the job losses, (iv) the decreasing of the foreign direct investment, (v) the high degree of corruption, (vi) the lack of cooperation with the International Monetary Fund (IMF) and (vii) many other social problems. Besides, the Republic of Serbia is currently engaged in the process of joining the European Union (EU) as a full membership, what is adding strong constraints on policy choices. The Serbian Government has a very limited financial capacity to assist citizens in risky situations, and therefore, it is not able to provide the necessary help to the jeopardized populations.

3.2.2 The Political Institutions

The legal framework to regulate the domains of safety and risk management are based on the Constitution of the Republic of Serbia [11] and European directives that have been already accepted in EU. The Serbian Constitution regulates and provides:

- Sovereignty, independence, territorial integrity and security of the Republic of Serbia, its international position and its relations with other countries and international organizations;
- Territorial organization of the Republic of Serbia (system of local self-government);
- Defence and security of the Republic of Serbia and its citizens (measures in case of emergencies);
- Sustainable development which includes a system of protection and enhancement of the environment, protection and enhancement of flora and fauna, industrial production, traffic and transport of weapons, poisonous, combustible, explosive, radioactive and other dangerous substances;
- Development of the Republic of Serbia which includes policies and measures for encouraging equal development of individual parts of the Republic of Serbia, the development of underdeveloped areas, organisation and utilization of a space, scientific and technological development.

The Constitution of the Republic of Serbia provides the right of citizens for province autonomy and local self-government, realised directly or through their freely elected representatives.

3.2.3 Organisation of the Competent Services

The current process of risk management can be discussed by presenting the Serbian emergency organisation. But previously, it is important to keep in mind that the national organisation to manage risks and emergency situations went through a period of great turbulence from 1994 till 2009. Indeed, the legal framework for disaster response had been created for a country and political system that no longer existed. That opened a space for other government structures to promote their responsibility for various aspect of crisis response and stepped into the vacuum that was created [12].

In 2009, the Sector for Emergency Situations (SEM) (Fig. 3.2) was established within the Ministry of Interior, based on the new law on Emergency Situations [13]. This new organisation has integrated the existing resources of the protection service, the rescue service and the reaction service in emergency situations. Its central mission is the protection and rescue of citizens, material and cultural goods and the environment in emergency situations.

The functionality of the Sector is supported by the twenty-seven local self-governments. The sector consists of:

- The department for prevention;
- The department for fire and rescue units;
- The department for Risk Management;

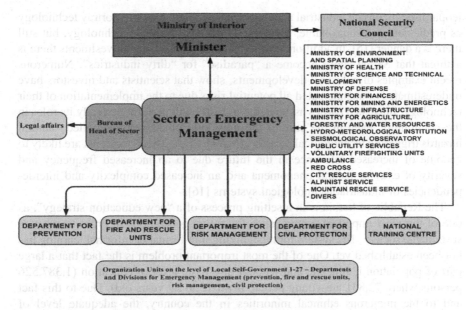

Fig. 3.2 Flow diagram describing the organization of the Sector for emergency management (SEM[2]) in charge of the process of risk management and risk communication

- The department for Civil Protection,
- The National Training Centre.

The role of the National Training Centre (NTC) is to train people involved in the management of emergency situations. After numerous emergencies, the experience feedback leads us to arise questions about the challenges and perspectives of the skills transmitted by NTC. More precisely, skills concerning fire fighting and rescue operations within the established emergency management system in the country, as well as others emergency services, like police and health workforce. The arising questions are:

1. How to ensure that their personnel have the necessary financial resources to perform tasks and missions assigned to them?
2. Which new knowledge and skills would they need to have in this extremely complex system?
3. How their education and training are supposed to be organized [14]?
4. How to improve the education and awareness of all participants involved in the risk management (the risk culture)?

Risk management in Serbia attracted great attention during numerous emergencies leading to significant consequences on population. The severity of these consequences is such that it is necessary to develop long term plans to remove them. Speaking about anthropogenic risks, the official representatives of the Government and experts from industry have wanted to reassure the public by promoting an idealised image of technology, in spite of numerous cases in which population was jeopardised by unsafe industrial activities. Stakeholders prefer to portray technology as predictable and controllable, speaking as the best available technology, but still there is a danger [15]. In addition, trying to attract foreign direct investments there is a threat that Serbia could become a "paradise" for "dirty industries". Numerous recent examples of industrial developments, show that scientists and investors have underestimated or unanticipated all potential risks due to the implementation of their technology. Interested parties assume that industry will perform perfectly its roles in the community, even in the case of the new kind of disasters so called "natural hazards triggering technological disasters (NATECH)". These accidents are likely to become of increased relevance in the future due to an increased frequency and severity of extreme natural phenomena and an increased complexity and interdependencies of industrial technological systems [16].

The Republic of Serbia is in a setting process of a "new education strategy", as one of the most important prerequisites for future generations in order to create a sustainable society. The system for recognizing non-formal or informal learning has not been established yet. One of the most important problem is the fact that a large part of population is illiterate and without any basic level of education (1,387,526 persons where 72,831 are young people from 15 to 29 years old). Due to this fact and to the numerous ethnical minorities in the country, the adequate level of security and safety in emergencies is difficult to achieve. The study of risk management in Serbia is new and largely theoretical. The academic community is looking for new ways to teach interdisciplinary approaches, experiential knowledge and best practices into its undergraduate and graduate curriculum. The purpose is to be closer of the problems emerging in the real-life, and to provide innovative solutions in the process of risk management.

Wider discussions about this issue is necessary as a platform for networking, exchanging of new ideas and approaches, transferring the knowledge and presenting best practices in risk management. In reality this discussion is still missing. On Serbian market, only few private companies offer services in the risk management area. As responses on various risks, they propose to make emergency plans, to provide trainings to acquire skills in communication during risky situations, etc. In many organizations (public or private), specific jobs of public relations were created to fulfill, among others, the risk communication missions.

The experience feedback concerning lacks and failures in risk communication has pointed the area of environmental risks as being the worst. Few positive steps suggested that establishment of the contemporary concept of risk management is possible, like in a process of acceptation of the Social Responsible Strategy [17]. Serbia has actively participated in the United Nation South Eastern Europe Disaster Risk Mitigation Adaptation Program (SEEDRMAP) and in numerous regional and international efforts to reduce existing risks.[1] The National Strategy of Protection

---

[1] The Global Facility for Disaster Reduction and Recovery (GFDRR) is a major initiative launched in September 2006. Within the context of the GFDRR the World Bank and UN/ISDR secretariat have initiated a South Eastern Europe Disaster Risk Mitigation and Adaptation Program (SEEDRMAP).

and Rescue that has been adopted [18], has to develop a multi-year programme which include research priorities in the areas of risk management and risk communication.

The role of media is an important factor for engaging and forcing stakeholders in real reforms in risk management and risk communication based on scientific approaches and on long term policy goals. From this point of view, a very important step was made as a part of the European Union project named: Development of strategic Planning and improvement of horizontal communication within the Ministry of Interior of Serbia.[2] As a result from this project, the Ministry of Interior has adopted and generalized the proposed communication strategy for it to become the basic guideline to be applied to develop communication in any area. The internal communication (including horizontal communication) has been highlighted to be one of the priorities. The transparency and publicity of the works are ensured through continual information of the public on the current activities and achieved results, through the work of the Media Cooperation Bureau and Bureau for Information of Public Importance, the publication of bulletins, etc.

Another important result from this project is the proposition to develop the dialogue with citizens according to two ways[3]: the internal and external target groups. The internal group comprises the Ministry staff. The external group consists of:

- The media—domestic and foreign;
- The citizens and non-government organizations;
- International Community—international organizations and embassies; and
- State institutions.

Despite the implementation of the new communication strategy described above, the whole process is not yet operational at the local level. Furthermore, according to the Peter Sandman description, the risk communication process at the local level can be qualified as the "first stage of risk communication", which consists simply to ignore the public. Despite the new legal framework, the risk communication in Serbia looks like the "pre-risk-communication stage, prevalent in the United States until about 1985" [19]. In this context, stakeholders have repeatedly shown that experts often disagree with people without scientific expertise (layperson), concerning the level and acceptability of a specific category of risks. Such disagreements caused lack of trust, differences in beliefs, gaps or errors in understanding risks, and differences in attention devoted to the risk management processes.

The capacities for a society to reach an adequate level of risk communication depend on the wealth and on the education of the population. The current Serbian society is not able to provide these necessary preconditions. The local self-governments are in the difficult position to make efforts both, for providing adequate safety on its territory

---

[2] It is a part of ongoing project financed by European Union Twinning number SR11 IBJH 01.

[3] Two-way communication in this document is defined as communication that has a goal to encourage open, straightforward, constructive dialogue, and facilitate the accessibility to information, influence, and engagement, listening to and consulting with internal and external user.

and for elaborating long-term sustainable development plans. Dutton suggested that the concepts, beliefs, assumptions and cause–effect understandings of strategist, determine how strategic issues will be framed [20]. Serbian policy makers have to show increased interest for such reflexions and they should include cognitive science advances in the processes of strategic decision-making in general and in the area of the risk communication in particular. It is a fundamental issue that could provide numerous possibilities to achieve common goals in the fields of risk and safety.

3.3 The Importance of the Risk Communication at the Local Level

The consequences caused by the last major accidents throughout the world can be considered to be more and more catastrophic. The strengthening of national capacities and specialized agencies, which has been recently implemented might be insufficient to face such situations requiring immediate emergency response at the front lines. The expansion of the local government roles within a coherent overall organization could be one of the accurate responses to this problem. According to the deep specific knowledge from the county they have to manage, local govern-ment roles could be expanded in the risk communication domain and in the implementation of local resilient organizations.

Strengthening the local resilience is a topic of paramount interest in the world as proved by the United Nation campaign: "Making Cities Resilient: My City is getting ready!". This campaign launched in 2010 addresses issues of local gover-nance and urban risks. It is entered in its second phase (2012−2015) and a large number of Serbian cities are included.

Most results from researches in communication acknowledge the importance of thinking communication as an exchange rather than a one–way process. The Serbian experience feedback has shown that a centralized communication (top–down pro-cess) is not adequate. A complementary bottom-up process is needed, followed with appropriate planning, prevention and systematic risk management at the local level.

Local self-governments' tasks include (i) the organization of various social, procedural, and economic activities, which achieve the consensus among various interest groups, and (ii) the decision-making process which contributes to build a safer community. There are numerous laws regulating the rights and responsibilities of local government as part of executive power based on territorial-administrative division of the Republic of Serbia. The most important is the Law of Local Self Government. The local self-governments are facing large regional development inequalities due to unfavourable demographic trends, high regional unemployment, devastation of industry, lack of infrastructure and insufficiently developed institu-tional framework. Having in mind that official national statistics have classified cities and municipalities in three different groups due to their level of development (developed, undeveloped, and devastated), it is clear that they are not in an equal position regarding response in risky situation.

By the Law on financing local authorities, which was in effect in 2006, it was expected that transfers from the national budget to the local authorities would be exactly 1.7 % of the Gross Domestic Products (GDP). Because of economic crisis, these transfers were reduced twice in 2009 and in 2010.It was the reason why mayors of more than fifty towns in Serbia handed the Government of Serbia a petition requesting the return of the full transfers from the budget of the Republic to the local authorities in 2011 [21]. These financial critical situations have raised the need for a new Law about changing and adapting the financing of the Serbian cities and municipalities [22]. The burden of sanitation and alleviation of risk conse-quences is usually left to the local authorities that are faced with the lack of skilled workforce and financial means. The local self-governments have the obligation to provide security, but they do not have the necessary preconditions to do it appropriately.

These numerous obstacles have initiated a new institutional approach in the management of risky situations at the local level. For these activities, Serbia needed international help from many various organizations. The response in emergency at the local level was one among very successful international programs. The corner stone of this program (Preparedness, Planning and Economic Security Program: PPES) was the risk management at local level and the role of citizen in emergen-cies, as well as daily activities in municipalities.[4] The PPES was the unique pro-vider of non-formal adult training programs in that field. It was filling the lack of response to the needs of training of the country for over more than twenty years, until the integrated disaster management system was established within the Ministry of Interior [23].

The program started in 2006, and the activities of its Component oriented towards strengthening capacities in disaster management at the local level were performed in 80 municipalities up to early 2011. More than 1,000 participants between 2007 and 2010 were attending the 2 days long workshop. The training was a combination of presentations, discussions, group works and debriefing sessions upon gained results.

Each workshop was evaluated by means of a questionnaire, where questions were scored from 1 to 5 (5 for the best score). For each training session (workshop), participants were grouped by two or three municipalities, what was representing an average number of 17 persons. A synthesis of the responses obtained from twenty out of forty trained municipalities in respective period (2008–2010), is presented below (Table 3.2) as an illustration of these evaluations.

---

[4]   The Preparedness, Planning and Economic Security Programs (PPES) are being implemented by DAI in Serbia, and organized in two components: Preparedness and planning (PP) and Economic Security (ES).

Table 3.2 Results gain after analysis of workshop objectives from participants divided in groups

Workshop	W1	W2	W3	W4	W5	W6	W7	Average score/ objective
Objective								
1. Identifying the role of local self government in emergency situations management	4.68	4.59	4.54	4.50	4.67	4.80	4.21	4.57
2. Identifying the role of public in emergency situations management at the local level	4.42	4.28	4.46	4.25	4.55	4.73	4.05	4.35
3. Defining the process of risk management	4.68	4.50	4.54	4.31	4.67	4.71	4.32	4.50
Average score/workshop	4.59	4.45	4.51	4.35	4.63	4.74	4.19	4.47

An important result emerging from this assessment is the difficulty encountered by participants to accept the paradigm shift in the way to consider the public. As a result, participators gave the lowest score at the objective regarding *the identification of the role of the public in emergency situation management at the local level* (at the time this concept was not yet developed in the country). The reason for these results comes from decades of application of the risk communication concept developed in the communist countries, which was entirely based on command and obedience of stakeholders in charge for mitigation, and by neglecting the role of the public. Therefore, population has "generally the impression that they will be the sacrificial lambs and that the self-government and its bureaucracy could not be trustworthy to do the right thing". That is a reason why almost all authors on the topic of risk communication highlight the importance of trust [24].

Despite the modern theory in risk communication, numerous examples have been recorded in Serbia where it failed in the last few years. One of the most representative examples is the explosion of a chemical factory near Belgrade. During this event, emergency services and public health services did not provide timely and accurate information for citizens. The most remarkable failure comes from the Serbian Parliament. A minister of the Government announced during the Parliament session, how "at that moment the police was escorting the transport of nuclear hazardous waste from Serbia to Russia". The whole action, which was qualified as top secret and which was organized by Serbian and international security services, has been so revealed to the public. With this statement, the minister caused civil unrest in the city where the waste had to be stored before transportation.

It appears that the implementation of the risk communication process at the local level is not yet taking in account the contemporary social and cognitive theories. Discussions and reflections about the place of the risk communication are neglected in Serbia because of the cultural heritage of the past habits in which old concepts such as "working *for* people" are still in use despite of new concepts such as "working *with* people". The current risk communication approach has its own shortcomings. Policy makers and local authorities do not still provide opportunity for real debate. They are focused on technical factors and they are neglecting social and political factors. In all activities, they did not perform any research to identify the target audience, as well as the pertinent tools and methods in order to communicate effectively and efficiently. In communities, there is not any leaflet, seminar, meeting, public report or media explanation about the intentions and the explanations concerning the responses to the existing risks. The human factors like age categories, cultural groups, social classes, life styles, education levels or gender are completely neglected. Furthermore, nothing is undertaken to remove obstacles that hinder an effective communication or to analyse the way to minimise these obstacles in the future.

There is an obvious need for stakeholders to take in account the progress towards the implementation of a more interactive risk communication. A variety of social science theories and models pertain to risk communication and how people think, reason, act, and make right choices [25]. Local authorities should consider these new theories and models as their strategic goals. Research works developed by Brewer [26] and Zhang [27] address issues that could help authorities to improve risk communication according to expected goals. Their models explain how information flow has to be, and how it will impact the judgement of the addressees.

Noel T. Brewer pointed in [26] that "risk communication is successful to the extent that it contributes to the outcomes its sponsor desires". The single risk information disclosure is not sufficient to change behaviours or beliefs of stakeholders and population. It should be a permanent process of information sharing. A permanent process of information sharing based on transparent relations among participants in the process, which allows participants to start thinking about widely accepted beliefs, is the first step which will lead to behaviour changes (Fig. 3.2). In this process the main obstacle is the need for long period to implement planned actions, which have to be limited in scope and location. As that process needs comprehensive plan and significant financial means devoted for that purpose, policy makers are not so interested to conduct it. Changing behaviours in a society needs mutual efforts at the national and international level.

Risk communication is centred on the intentionality of the information source and the quality of the information. As risk communication concerns at least two agents, the information process within each agent can be described by the CED model (Fig. 3.3): "Cognition process, Evaluation process, Direction process".

Fig. 3.3 Three potential goals of risk communication [26]

In accordance with this model, information flows from the source agent to the receiver agent, which in turn will be the new source agent. "In the cognition process the agent needs to recognize the fact, in the evaluation process it need to conduct the value judgement, and in the direction process the agent makes a corresponding decision" [27].

Hence, some authors have pointed micro/macro relations between cognition science and communication, and policy makers have to find a way to apply these new knowledge. The role of cognition and culture is fundamental in enhancing adaptive capacity of risk communication at local level. Among all stakeholders in the process of risk communications at local level, it is obvious that human factor researchers and practitioners including cognitive psychologists, and industrial engineers should work hand in hand with government officials and employees in the SEM.

Essentially the decision makers have to learn the lessons from the past problems and draw analogies between the causes and solutions for the current problems.

3.4 A Case Study About Hazardous Waste Policy

The effects of any kind of risks are amplified by the work of local self-government and society at whole. The presented situation confirms the theory of Harry Otway about one of the two kinds of risk communication: the one "used to persuade people to accept policies or technologies and their associated risks; which in essence encourages passive compliance with the intentions of those providing the information, and so on it is fundamentally manipulative" [28].[5] The case study is about an issue that has been seen similarly in all over the world. One of the most remarkable similar case is recorded by Sharon Beder and Michael Shortland in Australia where much of the interest in risk communication has come from government officials and others who have experienced difficulties in sitting hazardous waste facilities [29]. The role of adequate risk communication in all phases of the implementation of a waste management policy within a community is of paramount importance.

The Serbian case study, here presented, is a remarkable example of inadequate communication with citizens of the different cities in which hazardous waste was supposed to be stored. Some events within communities connected with this policy had attracted attention of public and experts from the fields of risk management and risk communication of the country and in the international community. The management of this case had caused significant consequences in the international relations.

---

[5] The second kind of risk communication according to Otway is a more ideal form, which aim is fulfilling the information needs of the audience so as to enable them to make their own decisions.

Hazardous waste disposal had caused the greatest anxiety among the Serbian citizen. From the 1.8 million of tones of waste annually generated by industry, one third are classified as hazardous waste. There is no permanent storage area for hazardous waste on the territory of the Republic of Serbia and the environment is affected in several ways accordingly [30]. The establishment of management systems for hazardous waste includes the establishment of adequate collection and transportation of hazardous waste, as well as the construction of [31]:

- Regional storages where hazardous waste should be kept for treatment (five such storages are planned);
- Treatment plants for hazardous waste (physicochemical treatment) within the very heart of factories, and
- Two incinerators for a thermal treatment of the hazardous waste, as well as the hazardous landfill.

The National Waste Management Strategy (NWMS) [32] had been the core document in the creation process of the necessary conditions for a rational and sustainable waste management. According to requirements of the NWMS, integrated waste management for a group of municipalities (in a region or a country) presents the only adequate waste management solution. After few years, it became necessary to update NWMS. It has been changed in 2010, and has been extended for the period 2010–2019. However, one of the key obstacles to the achievement of NWMS goals was the lack of waste management plan at regional and local levels.

The plan has been developed based on a legal framework, but its implementation has been faced with so many unexpected difficulties. Citizens started protests against building hazardous waste facility, first in the town named Cicevac, in Central Serbia, and then in Vojvodina and Kovacica. Those protests have caused delays in the implementation of accepted obligations in environmental policy, which is based on a signed contract in an IPA project entitled: "Technical assistance for hazardous waste management facility". The project was of national importance for the management of hazardous waste. Before those events the Government had tasked the Ministry of Environment and Spatial Planning to carry out the project "Construction of the physical-chemical treatment of waste". The Ministry has to work closely in cooperation with the Ministry of Defence, to prepare an agreement on mutual cooperation in order to determine the final location for the construction of the physicochemical treatment plants of waste, and to specify the way to transfer the right of use of the land. The two Ministries have actively participated in activities of interest to find a site for construction. It was expected that the realization of this project phase would start in 2011. The two ministries have only performed technical and political decisions without including any deeper analyses about the local communities in which these facilities will be placed.

After many misunderstandings and numerous citizen protests, in February 2012 Serbian authorities admitted to the European partners that the project is delayed because of public resistance in almost all sites where they plan to locate any of those facilities (Dobricevo, Drenovac, Cicevac and Kovacica). They were not able to find new locations that would fulfil all criteria defined in the project. The central bureaucracy made the mistake to ignore any support from environmental and community interest groups. In all cases they were cooperating only with local councils, which expressed responses on the basis of received information, but without consulting a broader population about these questions. The bad conduct of this highly sensitive project has obviously provoked political reactions from the opposition that has found a way to strongly reject this plan. The public reaction has been also hostile and the Council, which had first supported this project, has withdrawn its support for fear of losing future elections.

Following these events, and given the inability of the Serbian authorities to fulfil its obligations, the European Union has suspended the project. Stakeholders did not know the conditions in local communities and during these events they have never taken into account their prior attitudes and actions, neither their prior beliefs. They omitted to establish trust, which is a common thread in all risk communication strategies, and they did not consider the public as a full partner. It is well known that the public and members of social groups perceive risks differently from people (scientists and experts) who develop risk assessment studies. It is difficult to understand why authorities have neglected this fact. Population cannot be ignored in such areas where hazardous waste disposal facilities have to be built. Local authorities must undertake direct actions, they must respect and take in account the cultural and educational level of the population and more particularly about risk issues. Finally, permanent efforts have to be done to increase the number of people who can communicate effectively about risks.

Another dimension of such conflict situations is the role played by the media that report about conflicts in a way easily understandable by everyone, and sometimes simplifying the complexity of such real situations: the conflict between local authorities doing its best to deal with hazardous wastes, and local residents expressing the NIMBY syndrome (Not in My Back Yard). The trust dimension, which is too often forgotten in spite of its paramount importance in the problem of acceptability of a new technology, is the need for the population to trust the people who will construct, operate and regulate this technology. Having all above in mind, it is questionable why it has been decided to locate these plants in the poor and devastated communities. Such decisions raise the questions of equity and hazardous waste management. Reflecting similar concerns, in 1994 President Clinton issued Executive order 12,898, which calls on federal agencies to develop strategies to ensure "environmental justice".

There is a hope that Serbian authorities in the future could provide scientific data to perform researches about communities where waste facilities have to be built (e.g. the number of households with incomes below the poverty line). Indeed, residents from devastated and undeveloped communities might have more difficulties to access to the political system and therefore they might have less ability to oppose the building of new hazardous facilities than residents from developed communities. An additional research question could be to know if these factors (poverty, under-development, etc.) could influence manager decisions to determine the location of the facilities. In consequences, these facilities are they disproportionately located in minority communities, or in poor multiethnic regions (like Kovacica)?

This case study has confirmed that there were not any kind of communication between population exposed to "risks" and their own representatives. This lack of communication has turned the debates into a fighting match and has led the population to reject the project. The beginnings of a communication could have given rise to initial discussions on the issues of advantages and disadvantages of such a project of national interest. People who have to live around hazardous waste facilities are worried about it, even if experts believe the risks are slight. It is well known that when people are upset, angry, under high stress, involved in conflict or when they are very concerned about something, they often have difficulties to process information in a rational way. It is particularly important to consider this fact when a sensitive message has to be transmitted and well received.

In the case discussed as well as in similar cases, risk communication do not have to be a question of manipulating lay people to lead them in the desired direction, but to provide them with the necessary information to comprehend and evaluate the situation themselves [33].

3.5 Opportunities for Improvement of Risk Communication

"Communication has always been, if not the heartbeat, the circulatory system of science" [34]. Risk communication in global community has become a concept that is strongly marketed by specific interest groups and used instrumentally to achieve particular ends. The actual situation shows that Serbia suffers from a considerable lag in its risk communication policy compared to others developed countries. In the United States, the Environmental Protection Agency (EPA) has stated in 1986: "on the national level we will build risk communication into regulatory policy whenever possible" [35].

The Serbian policy makers must be guided by these cases and they must analyse them as experience feedbacks to take them in account in the elaboration of new policies. Authorities have to move forward and in the near future they have to correct their gaps and failures within a methodological framework in accordance with the five steps of the risk communication process: intentionality, content, audience, source, and flow. As the communication of information about risks usually occurs within a context of fear and uncertainty, its fundamental goal is to provide meaningful, relevant and accurate information, in clear and understandable terms targeted to a specific audience. It may not resolve all differences among all parties involved in process, but may lead to a better understanding of it. The other important question could be the acceptability of any risk, which is deeply connected to perceptions of fairness and justice. It may also lead to more widely understood and accepted risk management decisions, especially in poor and minority communities, and in undeveloped regions. Indeed, it could be a question of basic human rights and of environmental justice if hazardous waste facility will be settled in minority and undeveloped regions.

In fact the case presented above showed the requirement of on-going interaction, not only between regulatory authority and the regulated entity, the community and its self-government, but also communication among others interested parties. It generally involves adoption, compromise and negotiation during all process. Local communities and the general public reacted in case of limited, false, or inadequate information. The policy makers have to be aware of the possibility of antagonisms between the techno-sphere (the expert culture) and the demo-sphere (the popular culture) and they have to try to avoid them all the time.

Some of the researches in the risk perception area reinforce the conception that rationality and democracy are antagonistic to one another. As Serbia is evaluated like unconsolidated democracy it is important to address the role of media in risk communication. Stakeholders have to have in mind the current state of media freedom. In the last report about media freedom in 2010 Serbia is classified in a group of countries (36.5 % of the 197 countries and territories) where media are partly free, and gains 72 places [36]. The media has not played a particularly important role in mediating the processes of risk communication apart from the reader's letters chronicle in newspapers and debates between protagonists broadcasted over the electronic media. The communication with affected communities and interested groups tends to take place in community meetings and conferences, through correspondences, brochures and other publications, and by all the means that provide an active role for each individual to express its opinion.

The focus for most people studying risk communication has been on the ability of the communicator to instil trust in the communication. The task of Serbian education system is to educate communicators, but the great obstacles are the lack of scientific and academic programs about risk management and risk communication. It is not enough to address the public role in laws while otherwise its "right to know" and its participation in important decisions or activities are limited in many cases and due to many additional reasons. One important document adopted by the Parliament is the National Strategy for Protection and Rescue in Emergency Situations [37], but there is a lot of space to improve its implementation. The five strategic areas of the National Strategy are:

- To ensure that disaster risk reductions becomes a national and local priority with a strong institutional basis for its implementation;
- To identify, assess and monitor disaster risks and to enhance early warning;
- To use knowledge, innovation and education to build a culture of safety and resilience at all levels;
- To reduce the underlying risk factors, and
- To strengthen disaster preparedness or disaster response at all levels.

The strategy is based on priorities defined in Hyogo Framework for Action (HFA). The third priority about the use of "knowledge, innovation, and education to build a culture of safety and resilience at all levels", comprises four main aspects that can be summarized as follows: (i) risk information sharing, (ii) integration of disaster risk reduction (DRR) into curriculum/education, (iii) development of multi-risk assessment and cost benefit analysis, and (iv) public awareness. The Hyogo

Framework for Action 2005–2015 [38] primarily emphasizes the significance of national and local governments in reduction of risks. As a result, they need to create institutional and legal frameworks, to provide academic resources and to ensure the participation of communities. It encompasses:

- Relevant information on disasters available and accessible at all levels and to all stakeholders (through networks and development of information sharing systems);
- School curricula, educational materials, and relevant trainings concerning concepts and practices of the risk reduction and recovery;
- Development and strengthening of the research methods and tools for multi-risk assessments and cost benefit analysis;
- Development of a countrywide public awareness strategy to stimulate a culture of disaster resilience, with outreach to urban and rural communities.

At any level (local, regional and national), the most important for risk managers is to communicate about risks, and to deliver the message in a competent manner. Peter Sandman, the "father of modern risk management" delineates Four Kinds of Risk Communication depending on the public perceptions about "hazard" (the technical component of risk) and "outrage" (the principal determinant of perceived hazard) [39].

In the future, additional tasks for policymakers and academia should also be to determine what activities the Serbian society could undertake for preparing students for emergencies, and what could be the core of competencies they should have. The person who is in charge to communicate has to know how to translate technical and scientific concepts into understandable messages. In general, recipients of a threat information must be in capacity to receive information, to understand information, to understand that the message apply to them, to understand that they are at risk if they do not take protective actions, to decide that they need to act according to the information, to decide which actions have to be taken and that they must be able to take action. The mistrust in risk management authorities and the public demands for the right to participate as a full partner in all phases of risk assessment and risk management are absolutely justified.

In Serbian cities and municipalities there are obvious lack of design and implementation of new systems and methodological tools that can help to solve problems in the domain of the risks. Policy makers have to create adequate guidelines on good practices for risk communication, based on the existing literature published by different experts and authorities all over the world. Among this literature, the U.S. Environmental Protection Agency has published seven cardinal rules of risk communication, which has been widely quoted and used in practice. Seven Cardinal Rules of Risk Communication by Vincent T. Covello and Frederick W. Allen became an alphabet in work of emergency services in USA. Those rules which has to be implemented are:

1. Accept and involve the public as a partner;
2. Plan carefully and evaluate your efforts;

3. Listen to the public's specific concerns;
4. Be honest, frank, and open;
5. Work with other credible sources;
6. Meet the needs of the media and
7. Speak clearly and with compassion.

From our point of view, another concept, which is recognized as very useful in a process of improvement of risk communication, is the "Crisis and Emergency Risk Communication" concept (CERC). It could be upgraded previously to be implemented in Serbian conditions. The goals of this concept are to help people cope, to empower decision-making, and to begin to rebuild a sense of normalcy in the life of the endangered population. CERC is a relatively new scientific concept of communication and unknown like many others among policy makers and academic community in Serbia. This concept tries explaining the psychology of a crisis and its impact, providing tools to prepare for and respond to the communication challenges that occur in times of emergency. It has emerged as a new field of communication recognized by academia and broader scientific community, and has first practical implications in USA in the work of different actors. CERC is recognized as a conceptual overview, a synchronization of basic principles accepted from crisis communication, a management communication, a risk communication and finally a crisis and emergency communication. "Crisis and Emergency Risk Communication" is taught in 22 universities in the USA and is being diffused internationally. The most influent international organizations have adopted CERC principles like The Pan American Health Organization (PAHO), the World Health Organization (WHO), and the North Atlantic Treaty Organization (NATO). CERC is vital because it helps the public to respond to crises, it reduces the likelihood of rumours and misinformation and it demonstrates a good leadership.

The education for the future responders depends on quality of programs provided in Serbia and the competent institutions have to think about it. Public stakeholders in charge of the crisis communication have to accept the six basic principles of CERC [40]:

1. Be First—You have to provide information from authority as soon as it is possible.
2. Be Right—Give facts in increments.
3. Be Credible—Tell the truth.
4. Express Empathy—Acknowledge in words what people are feeling—it builds trust.
5. Promote Action—Give people things to do.
6. Show Respect—Treat people the way you want to be treated—the way you want your loved ones treated—always—even when hard decisions must be communicated.

Serbian policy makers in a process of seeking full membership in European Union have to confirm that they understand and that they are able to apply the modern concepts of strategic management in a modern society. They must be aware of the increased importance of the cognition in strategic decision-making. They have to understand relations between cognition and risks and therefore relations between cognition and risk communication. In the process of risk communication, they have to understand the importance of knowing the target group in taking into account the prior attitudes and actions, beliefs and decision circumstances. Risk communication can change the risk mental models by correcting errors, shifting emphases, filling in gaps, and providing details to specify vague or general beliefs. The advantages to design risk messages based on an approach centred on mental models are developed by Bostrom [41] and Morgan [42]. This approach addressed the importance of the content design by taking into account the decisions people have to make, the potential message, the recipients' prior mental models, and what experts know about the physical hazards underlying the risk and about the risk reducing measures.

The experts from emergency services have to coordinate their work with researchers from cognitive science in the planning or risk management and risk communication. The contextual control model (CCOM) of cognition could be appropriate to avoid current difficulties in the risk communication process [43]. In recent researches, the contextual control model has been enriched with some additional explicit factors relevant for risk communication. This model (Fig. 3.4) presents that agents receives information from the source of information and trough the process of cognition, evaluation and direction.

In Ekberg works is precisely explained the different model which purpose is that it puts the decission maker into the context in which it has to perform. This model (Fig.3.5) could be very useful for future use in Serbia during risky sitsuations.

Contemporary risk management policies have to be consistent with three basic elements: (i) goal setting, (ii) information gathering and (iii) actions to influence behaviours (or risk mitigation). Adequate risk communication is the most important element in achievement of behavioural changes. Therefore, the improvement of risk communication needs a stakeholder's strong commitment to work towards a more coherent approach of the risk communication, as a specific part of risk policy. Development of a coherent approach to conduct risk communication actions should be subject to a range of new regulations devoted for this purpose. Stakeholders, which are presented in Table 3.3 (fourth column, due to their actions, cannot afford

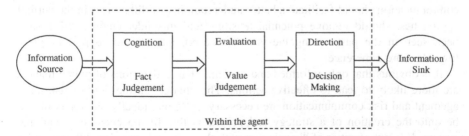

Fig. 3.4 The CED information process model (*Source* https://www.jsce.or.jp/library/open/proc/maglist2/00039/200511-no32/pdf/108.pdf)

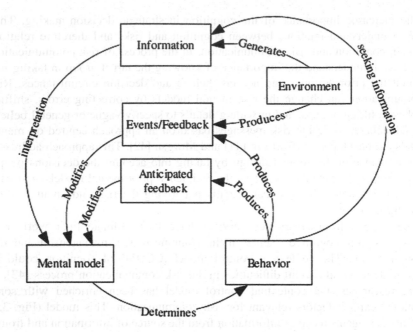

Fig. 3.5 Contextual system model of risk cognition (from Ekberg, 2007) [44]

to wait any longer and should create conditions for the required changes as early as it is possible. They have to communicate and coordinate their efforts to address a variety of themes which affect many aspects of risk communication issues in the areas of (i) education (formal and informal), (ii) budget planning (keeping in mind the budget shortfalls), (iii) implementing the already developed public-private partnership actions (characteristic for developed countries), and finally (iv) bridging the gap between science community, policy makers and civil society. All of these actions are compounded by a staggered timeline for change. There could not be further excuse for any partial approach of the different institutions. It is time for harmonized approaches concerning risk communication, within and between the different policy spheres of the Serbian society.

The implementation of a new system of responses (Table 3.3) in the area of risk communication should facilitate a better understanding of differences in the current approaches, should remove potential tensions and misunderstandings (that have been seen in the past among the various actors), and should support progress towards more coherence.

It is obvious that only informed and evidence-based decisions of policy makers are more likely to lead to effective actions. The changes in the field of risk management and risk communication are necessary in Serbia. Ideally, the focus should be onto the creation of a strategy and programs that fit the needs of a national system. It is not a matter of "keeping up with the Joneses", creating differences for the sake of differences, or rejecting a specific approach because somewhere else

Table 3.3 Proposal of a possible "system of responses" to improve risk communication as a part of risk management in the Republic of Serbia

Identified problem	Responses	Implementation	References/Authors
Part 1			
The low level of education in Serbia	Creation of a new national Strategy for education and development of the scientific research	Creation of new courses in the area of the risk communication in curriculum of educational institutions	Serbian Government; Ministry for Education, Science and Technological Development; Universities; faculties
Insufficient training of competent authorities to provide risk communication in practice	Providing training for persons in charge to communicate with citizens in accidents and various kind of emergencies	Applying some of the accepted methods of risk communication in international community like CERC or others	Ministry of Interior affairs; Sector of Emergency Management; National Training Center; NGO; Academic community; Ministry of Defense; Ministry of Health; Ministry of finance; Local self Government
Insufficient level of coordination and cooperation among the stakeholders. Inheritance from communist regime approach: "Work for people" instead of "work with people"	Changing awareness. Risk communication and risk management is a task for all, not a privilege for a few one who decide to speak in the name of all of stakeholders	Implementation of existing laws and international conventions in practice; Creation of standard operation procedures among all stakeholders; Involving media in all activities; Spreading responsibility at local level; Working on increasing awareness about teamwork	Serbian Government; International Community; European Union; Ministry of Interior affairs; Sector of Emergency Management; Ministry of Defense; Ministry of Health; Ministry for energy and environmental protection; Ministry of finance; Local self Government; media

(continued)

Table 3.3 (continued)

Identified problem	Responses	Implementation	References/Authors
Part 2			
Budget shortfalls in local self government, regional and national level. Greater participation of citizens and private sector in a budget creation at all level: municipality, regional and national	Competent authorities have to have in mind all kind of risks and due to that create the budget. The budget has to be devoted to the most dangerous risks based on science prediction not on perceived risks. The citizen's and private sector participation in the budget creation has to be providing through following a practice from democratic communities	Budget devoted to emergencies, risk communication and risk management have to be preciously accounted and proposed by experts in the areas of finance and disaster management together; It has to be banned that every year competent authorities merely renew the same budget without conducting any analysis of past or predicted financial means in emergencies	Serbian Government; Serbian Parliament; Competent Ministries, first of all the Ministry of Finance; Self governments; Public private sector; NGO; Media; Academic community
Wide gap between science community, policy makers and population	Bridging the gap between them increasing awareness with use of newly scientific researches in the area of risk communication, such as a linkage between cognitive science and risk communication and management	Providing financial means for scientific research among experts from emergency services with researchers from cognitive science. Participations of Serbian professors and high education institution in COST action ISCH 1,304 Expert Judgment Network: Bridging The Gap Between Scientific Uncertainty And Evidence-Based Decision Making, as well in other numerous projects in country and all over the European Union	Stakeholders in the area of emergency management at all levels in Serbia; Researchers from psychology, sociology and cognitive science area; European Union, International organizations such as United Nations, UNDP, UNEP, World Health Organization and World bank, and etc.; Citizens itself, NGO

someone has a similar one. It is a matter of making sure that, whatever approach is being considered, it has to fit the context into which it will be operating" [45].

3.6 Conclusion

In a global world, risk management and risk communication need an increasing amount of attention and a systematic approach. This is particularly important for the Republic of Serbia, which was, and still is exposed to social, economic, and political transitions in harsh circumstances.

Among stakeholders, the Sector for Emergency Management has the most important role in the risk communication area. The risk management and risk communication are equally important for any hierarchical level in Serbia, but it does not mean that all of them have the same conditions to achieving the common goals. The risk management system in Serbia on the local level has to be a part of special concern of policy makers, non-governmental organizations and civil society.

Serbia still lack of scientific and academic programs about risks management and about communication skills in general. Those weaknesses are recognized and they will be removed in the future by creating new curriculum at universities. On the same way, it will be avoided that practitioners only rely onto their personal attitude, their intuition, and the application of unproven best practices. Knowledge and innovation, education and availability of information, research, discussions, and training have to become top priorities in the fight against the risks.

The process of the risk communication should take better account the research advances in psychology than it is currently done. Choosing a goal for risk communication is the most needed for risk communication plan creation. Changing behaviour as a goal for risk communication requires knowing the action plan to perform. During that process policy makers have to understand that no single method of message delivery can be considered to be sufficient. They have to incorporate in new risk communication strategy the needs of target audience, and multi-facetted delivery method that could be effective at reaching the largest audience. The message processing depends on the cognitive characteristics and cognitive limitations of the message recipients, as well as the characteristics of the message (reliability, suitability, consistency, clarity, comprehensibility, etc.).

The case study about hazardous waste facility has served as colourful example for the current weakness and deficiency in the risk communication. Almost all theories on the topic of risk communication stress the importance of trust. The contradiction between official statements of reassurance and other less conscious statements of risk does nothing to reinforce trust in the self-government. Judgments are not always based on simple agreement or disagreement with policy statements, but are sometimes based on inferences from those statements as well, especially when people disagree with the specific content of what they've heard or read. The worst is the case when authorities lost the trust of their citizens, and make a gap between them and the interests and concerns of the public. In future, majority of inconvenient situations could be avoided by conducting the consultation process by some kind of "taskforce" before the decision is made. In the future, local population should no longer be excluded from the consultation process and stayed just as an observer, waiting how hazardous waste facilities will be built in its municipalities.

The literature provides much information in developing efficient risk information. In risk communication, individuals and institutions would like to exaggerate or underestimate risks for various reasons. To get a risk information more understandable, the developing process may benefit from risk communication theories, judgement and decision making, researches about mental models, etc. The adoption and implementation of the CERC model or similar communication models would be a chance for improving transparency, accountability and cross sectional coordination, and for implementing an effective two way communication especially in multiethnic region.

Risk communication must be revised in the future work of policy makers. It will be a first step of building an adequate risk communication based on reviewing past successes and failures. Fischhoff explained that a communication is adequate if: it contains the information needed for effective decision-making; users can access that information and they can comprehend what they access. Bostrom predicted in 2003 that "In the future, ubiquitous communications aids in the form of smart risk agents could tailor both the medium and the message, placing a risk in context for a specific person". At the present time Serbia is far from this concept, but a new era characterized by the implementation of interdisciplinary approaches and cognitive science to improve risk communication, will begin as soon as it is possible.

References

1. Bostrom, A. (2003). Future risk communication. *Futures, 35*(6), 553–573. www.sciencedirect.com/science/journal/00163287/35.
2. Beck, U. (2006). Living in the world risk society. *Economy and Society, 35*(3), 329–345.
3. Bailenson, J. N., & Fox, J. (2008). Cognitive science. In W. Donsbach (Ed.), *The International Encyclopedia of Communication, 2*(9), (pp. 548–551). Malden, MA. Wiley-Blackwell.
4. Plough, A., & Krimsky, S. H. (1987). The Emergence of risk communication studies: Social and political context. *Science, Technology, Human Values, 12*(3/4), 4–10. Sage Publications. Inc.
5. International Programme on Chemical Safety. (2004). IPCS risk assessment terminology. Harmonization Project Document No. 1. WHO document production services, Geneva, Switzerland. www.who.int/ipcs/methods/harmonization/areas/ipcterminologyparts1and2.pdf.
6. Sandman, P. (2003). Four kinds of risk communica-tion. http://www.petersandman.com.
7. Glik, D.C. (2007). Risk communication for public health emergencies. *Annual Review of Public Health, 28*,33–54. The Annual Review of Public Health http://publhealth.annualreviews.org.
8. European Environmental Agency. (2010). Technical Report/No 13/2010. Mapping the impacts of natural hazards and technological accidents in Europe. An overview of the last decade. Luxembourg (Luxembourg): Publications Office of the European Union; 2010. http://www.eea.europa.eu/publications/mapping-the-impacts-of-natural/mapping-the-impacts-of-the.pdf.
9. Global Risk Identification Program (GRIP). (2010). Serbia Assessment Report. For a purpose of regional program on disaster risk reduction (DRR) in South East Europe, sponsored by

WMO/UNDP Joint activities. http://www.gripweb.org/gripwebo/gripweb/sites/default/files/
Serbia%20assessment%20report%20from%20government%20input_SRBinput_0.pdf.
10. World Risk Report (2012). BündnisEntwicklungHilft, UNU, EHS Swiss RE, Weathering
    climate change: Insurance solutions for more resilient communities (2010).
11. Constitution of the Republic of Serbia ("Official Gazette of RS" No. 98/06).
12. Radovic, V., Vitale, K., & Tchounwou, P. B. (2012). Health facilities safety in natural
    disasters: experiences and challenges from South East Europe. *International Journal of
    Environmental Research and Public Health, 9,* 1677−1686.
13. Law on emergency situations (Official Gazette of RS, number 111/09).
14. Karabasil, D., & Radović, V. (2010). Is there a need to improve the fire service in the area of
    emergency management within the Republic of Serbia. In *Proceedings of International
    Scientific Conference, Fire Engineering.* October 5−6, Zvolen, Slovenia.
15. Radović, V., & Domazet, S. (2010). The role of the privatization process in Serbia as a
    function of jeopardizing the safety of citizens and the environment—drastic examples.
    Business Economics, 4(2), 151–169.
16. European Environmental Agency. (2010). Technical Report/No 13/2010. Mapping the impacts
    of natural hazards and technological accidents in Europe. An overview of the last decade.
    Luxembourg (Luxembourg): Publications Office of the European Union; 2010. http://www.
    eea.europa.eu/publications/mapping-the-impacts-of-natural/mapping-the-impacts-of-the.pdf.
17. Strategy of development and promotion of the social responsible business in the Republic of
    Serbia for the period from 2010 to 2015, Retrieved December 16, 2014, from http://www.
    srbija.gov.rs/vesti/dokumenti_sekcija.php?id=45678. In Serbian language, Official Gazette of
    the Republic of Serbia, No 55/05;71/05,corrections101/07 and 65/08.
18. National Strategy for Protection and Rescue in Emergency Situations, Official Gazette of the
    Republic of Serbia No 86/11 (2011).
19. Covelo, V., & Sandman, P. (2001). Risk communication: Evolution and revolution. In
    A. Wolbarst (Ed.), *Solutions to an environment in Peril* (pp. 164–178). Baltimore. MD: John
    Hopkins University Press.
20. Dutton, J., Fahey, L., & Narayanana, V. (1983). Toward understanding strategic issue
    diagnosis. *Startegic Manangement Journal, 4,* 307–323.
21. Radovic, V., & Komatina Petrovic, S. (2012). From failure to success: Serbian approach in
    mitigation of global climate change and extreme weather events. *Journal of Environmental
    Protection and Ecology, 4*(13), 2207−2214.
22. Law on Financing Local Self-Government, Official Gazette of the Republic of Serbia No. 47/11
    (2011).
23. Orlovic, L., & Radovic, V. (2011). Evaluation and practical implications of adult training
    activities: from the experiences of disaster management capacity development program in
    Serbia. In *Proceedings of the 14th International BASOPED Conference "Evaluation in
    Education in the Balkan Countriess".* Belgrade, Serbia.
24. O'Rordan, T., et al. (1989). Themes and tasks of risk communication. Report of an
    International Conference held at KRA Julich. *Risk Analysis, 9*(4), 514.
25. Science Panel on Interactive Communication and Health (SciPICH). (1999). In Thomas R.
    Eng & David H. Gustafson (Eds.), *Wired for Health and Well-Being: the Emergence of
    Interactive Health Communication.* Washington, DC: US Department of Health and Human
    Services, US Government Printing Office, April 1999. http://www.health.gov/scipich/pubs/
    finalreport.htm.
26. Brewer, N. (2011). Introduction to evidence based communication. In B. Fischhoff, N.
    Brewer, & J. Dowins, (Eds.), *Communicating Risks and Benefits: An Evidence-Based User's
    Guide* (Chapter 2). Retrieved from http://www.fda.gov/ScienceResearch/SpecialTopics/
    RiskCommunication/default *.htm.*
27. Zhang, Ch., Okada, N., & Matsuda, Y. Three types of risk communication patterns and
    corresponding models discerned from the analysis of actual disaster management process.
    www.jsce.or.jp/library/open/proc/maglist2/00039/200511-no32/pdf/108.pdf.
28. Otway, H. (1987). Experts. *Risk Communication and Democracy. Risk Analysis, 7*(2), 127.

29. Beder, S. H., & Shortland, M. (1992). Siting a hazardous waste facility: The tangled web of risk communication. *Public Understanding of Science, 1*(2), 139–160.
30. Prokic, D., & Mihajlov, A. (2012). Contaminated sites. Practice of solid waste management in a developing country (Serbia). *Environment Protection Engineering, 38*(1), 81–90.
31. Serovic, R. (2011). Hazardous Waste Management in Republic of Serbia. Ministry of Environment and Spatial Planning, Belgrade. http://www.twinning-hw.rs/wp-content/uploads.
32. Ministry for the Protection of Natural Resources and Environment (NWMS). (2003). *National Waste Management Strategy with EU.* Approximation Programme: Belgrade, Republic of Serbia.
33. Ekberg J. (2007). Risk Cognition Methodologies for Development of Mental Models of Risk Communication during Pandemic Influenza Outbreak. Master thesis in Cognitive Science. Linköping University, Sweden. Retrieved December 17, 2014 from http://www.diva-portal.org/smash/get/diva2:22977/FULLTEXT01pdf.
34. Lucky, R. (2000). The quickening of science communication. *Science, 289*(pp. 5477), 259 −264. http:www.sciencemag.org.
35. Thomas, L. (1986). Risk communication: Why we must talk about risk. *Environment, 28*(2).
36. Report Freedom of the Press. (2012). Global Press Freedom rankings. http://www.freedom house.org/sites/default/files/Global%20and%20Regional%20Press%20Freedom%20Rankings. pdf.
37. National Strategy for Protection and Rescue in Emergency Situations. (2011). *Official Gazette of the Republic of Serbia,* No. 86/11.
38. United Nation International Strategy for Disaster Reduction. (2005). *Hyogo framework for action 2005–2015: Building the resilience of nations and communities to disasters.* Retrieved January 12, 2014 from http://www.unisdr.org/files/1037_hyogoframeworkforactionenglish.pdf
39. Sandman, P. (2003). *Four kinds of risk communication.* http://www.petersandman.com.
40. Reynolds, B., Galdo J.H., & Sokler L. (2002). *Crisis and emergency risk communication.* Centers for Disease Control and Prevention. http://www.cdc.gov/communications/emergency/cerc.htm.
41. Morgan, M. G., Fischhoff, B., Bostrom, A., & Atman, C. J. (1992). Characterizing mental models of hazardous processes: a methodology and an application to radon. *Journal of Social Issues, 48*(4), 85–100.
42. Morgan, M.G., Fischhoff, B., Bostrom, A., & Atman, C.J. (2001). *Risk communication: a mental models approach.* Cambridge University Press.
43. Hollnagel, E., & Woods, D. A. (2005). *Joint cognitive systems: Foundations of cognitive systems engineering.* Boca Raton: FL, CRC Press.
44. Ekberg, J. (2007). Risk cognition methodologies for development of mental models of risk communication during pandemic influenza outbreak. *Master thesis in Cognitive Science.* Department of computer and information science Linköpings universitet http://urn.kb.se/resolve?urn=urn:nbn:se:liu:diva-8102.
45. Britton, N.R. (2004). *Higher education in emergency management: What is happening elsewhere.* Paper presented at the 7th Annual By-Invitation Emergency Management Higher Education Conference. National Emergency Management Training Center, Federal Emergency Management Agency. Department of Homeland Security. Emmitsburgh, Maryland. 8–10 June 2004.

Chapter 4
Metaphors and Risk Cognition in the Discourse on Food-Borne Diseases

Magdalena Bielenia-Grajewska

Abstract The aim of this contribution is to discuss the most important issues connected with communicating food-borne diseases. The investigation focuses on the role of symbolic language in informing stakeholders about food-borne crisis situations. To narrow the scope of the research, the study concentrates on metaphors and their role in risk cognition, especially in the face of information overload. This approach allows the author to study the metaphorical dimension of risk cognition as well as its dynamics connected with the necessity of a constant response to changing internal and external conditions. Theoretical investigations on metaphors in food, health and risk discourse are often supported by empirical analyses on the use of metaphors in communicating food-borne diseases. This study encompasses selected materials on food-borne diseases gathered from Italian online sources and covers the corpus of investigated verbal metaphors. The aim of this chapter is to show whether metaphors strengthen or weaken risk cognition, and how far they determine the risk communication of food-borne diseases.

4.1 Introduction

Modern times can be characterized by the influence of networks, information load, and various risks that determine their current shape. First of all, the reality of the twenty-first century can be examined from the perspective of a network society that can be defined as a *society whose social structure is made of networks powered by*

M. Bielenia-Grajewska (✉)
University of Gdansk, Gdansk, Poland
e-mail: magda.bielenia@gmail.com

M. Bielenia-Grajewska
Scuola Internazionale Superiore di Studi Avanzati, Trieste, Italy

© Springer-Verlag Berlin Heidelberg 2015
J.-M. Mercantini and C. Faucher (eds.), *Risk and Cognition*,
Intelligent Systems Reference Library 80, DOI 10.1007/978-3-662-45704-7_4

microelectronics-based information and communication technologies [1]. Moreover, both private and business spheres are determined by complex grids and lattices responsible for the contacts among human beings, as well as their relations with other entities [2]. Since nowadays there are fewer and fewer time or space barriers, distances of a geographical, political and cultural nature are becoming less and less visible [3]. Moreover, this borderlessness can be observed at both group and individual levels since companies are a part of the fluid reality [4] and, at the same time, individuals possess hybrid identities [5] that undergo changes, responding to alternations in their environment.

Secondly, the growing popularity of social media, the Internet and email correspondence makes many people suffer from *information overload* that takes place when the quantity of information is excessive [6, 7] or disruptive, owing to the vast number of impulses offered by various information sources that continuously require recipients' attention [8]. As a consequence, individuals may not understand some data, may have problems in judging whether the piece of news is reliable or may doubt that such information exists [9]. Since information overload is also connected with the necessity to select the required parts of information among the multitude of offered resources, it is important to use linguistic devices that enhance and foster data selection and comprehension. In addition, the price of information does not depend on its cost but on its value [10] and, consequently, linguistic tools are often used to enhance the merit of available data or make them more valuable in comparison with similar pieces of information.

The third element determining the performance of modern organizations and individuals is risk. Although it is said that people have to deal with the overproduction of risks in the modern reality [11], it should be mentioned that the twenty-first century is not more dangerous than the past epochs, but what has changed in our society is the level of knowledge and awareness of risks [12]. Consequently, in the economics of the third wave, workers that are needed should be thinking, critical, creative and ready to take risks [13]. Although there are various hazards present in modern reality, the risks that bother both individuals and organizations are those related to health. As Sontag [14] states, *"illness is the night-side of life, a more onerous citizenship. Everyone who is born holds dual citizenship, in the kingdom of the well and in the kingdom of the sick. Although we all prefer to use only the good passport, sooner or later each of us is obliged, at least for a spell, to identify ourselves as citizens of that other place"*. Thus, a disease is a part of one's existence and determines it to a smaller or greater extent. Although in most sources diseases are pictured in a negative way, illnesses are not only one's enemy but they can also be perceived as a friend since they show what one's organism lacks and what should be done to become healthy again [15]. Since diseases are one of the aspects shaping one's identity [16] and estimating risks raised by experts is a part of identity creation, built on response and condition [17], food-borne diseases studied from a discursive perspective may offer an interesting discussion on modern identity in risky environments.

4.2 Food-Borne Diseases from the Risk Perspective

Food-borne diseases are caused by the consumption of contaminated foods or beverages. Although there have been significant improvements in food safety, including the pasteurization of milk, safe canning and disinfection of water, food-borne poisoning still causes health problems for consumers [18]. Consequently, dealing with food-borne diseases continues to be of great importance, and such issues as food preparation, food storage and hygiene are frequent topics of scientific discussion [19]. It should be added that although the home environment is associated by most people with safety, security and relaxation, it is also a potent locus of possible infection, as food-borne diseases are often caused by poor hygiene in home kitchens and improper food preparation by consumers at home [20]. Domestic kitchens are also the place where consumers tend to take decisions on healthy nutrition since a person preparing food at home may face the risks of serving food with preservatives or offering potentially spoiled food, as a result of problems with storage [21].

As far as hazard typology is concerned, food-borne diseases follow similar risk characteristics to SARS (Severe Acute Respiratory Syndrome) or avian influenza. They can be characterized as *unknown risks* since (at least in the initial stages) virus origins and infection processes are often undetected. They additionally constitute *dread risks* since infections may be fatal, affect many people and remain untreated because no effective medication is available at a pandemic outbreak [22]. A substantial problem concerning food-borne illness is the issue of under-reporting. Even if seen by a doctor, the patients' stool samples are not taken and analyzed very quickly. Thus, it becomes very difficult to estimate the real number of people who suffer or have suffered from a particular disease, and reported cases constitute less than 10 % of all those infected [23]. Further aspects of risk communication related to the food industry are the credibility of information sources [24] and proper communicative strategies.

4.3 Communicating Food-Borne Diseases

Food-borne disease, together with food irradiation [25], genetically modified food [26], food recall [27], food contamination [28], food allergies [29] and food safety regulations [30] belong to very popular topics in communication on food risks.

It should be underlined that the perception of risks related to food-borne diseases is connected with different social notions. According to the *Health Belief Model*, the perception of threat is determined by such factors as age, gender, ethnicity, experience, education, knowledge and socio-economic status [31]. In addition, individuals are likely to take preventive actions if several determinants may be observed. They are as follows. *Perceived susceptibility* is when someone feels likely to suffer from a disease. *Perceived severity,* on the other hand, concerns an opinion that a disease is connected with serious effects on health. The third issue, *perceived benefits*, reflects a potential beneficial aspect of preventive actions. To add, preventive behavior is likely to take place when some benefits of health-oriented attitudes outweigh various expenses related to them [32]. Moreover, cultural differences shape the perception of illnesses. For example, some nations are more likely to view illnesses as more risky than other ones. In addition, individuals vary in attitude towards the credibility of information on risks provided online [33]. Moreover, geographical factors are important in risk perception. For example, people in New Zealand may estimate that their geographical location makes them less likely to become exposed to some diseases. It has been proved, however, that a geographical locus does not protect from diseases since e.g. H1N1 came to New Zealand in 2009 together with some returning participants of a school trip to Mexico and the USA [34]. Thus, information on a geographical scope of food risk situations determines its perception. In addition, when food poisoning has been detected in a distant country, a disease itself may be perceived as less risky for those located far away from a pandemic outbreak. The same applies to a victim group. For example, if a virus is supposed to attack a certain age group, representatives of other generations may feel safer. Furthermore, as [35] state, taking the optimistic bias effects into account, people underestimate food risk. As a consequence, individuals may think they are less prone to risks than other people. Moreover, risk cognition depends on such factors as an individual's world view, moods, emotions and immediate stimuli [36]. Thus, one's bad humor or stress related to other situations may result in an improper risk estimation. Moreover, people judge a disease as more risky if it is likely to cause serious health outcomes [37]. The following factors determine the perception of risk: an expected number of fatalities or losses, possible catastrophic potential, perceived properties of risk sources or risk situations and beliefs connected with risk determinants [38]. For example, applying a positive frame by showing a number of people saved makes individuals more likely to select health programs for combating diseases [39].

The aims of risk communication are as follows. First of all, its intention is to enhance the understanding of risks among different stakeholders. The second goal is to alter individuals' ordinary behaviors with the aim of reducing health hazards. The third issue concerns an increase of trust and credibility of institutions responsible for dealing with risks. The fourth aim concerns enhancing dialogue and solving conflicts [40]. Another issue important in risk communication is called *disaster fatigue* and it concerns situations when the general public is faced with information on different diseases [41]. Taking into account the multitude of data on health-related risks in the media, the aim of information creators is to use such linguistic resources to communicate food-borne diseases that draw stakeholders' attention to important issues. Their proper selection is important since the way food-borne diseases are pictured shapes an attitude towards them. For example, the events that are quickly perceived in one's mind are rated as more probable than the ones that require more effort as far as perception and comprehension are concerned [38]. To continue the discussion on comprehensibility of information on food-borne disease, it should be mentioned that proper representation makes unknown or novel scientific concepts understandable by diversified stakeholders [42]. Taking into account the linguistic dimension of risk communication, selected linguistic tools transform scientific information into the content that would be easily understood by the general public [43]. For example, humorous graphic images may be very helpful. Another important issue is a context itself [44]. What is more, a text is supposed to be simple, sentences should be short and coordinate sentences, together with very technical terms, should be avoided [45]. Moreover, such adjectives and adverbs as *highly, excessive, dramatically* or *extremely* stimulate the perception of risk. The same applies to such phrases as *well established, widely agreed upon*, and *widely acknowledged* that possess the idea of expertness in them, and consequently, make the reader believe in such statements [46]. Furthermore, numerical information may also increase or decrease risks in the eyes of readers, taking into account numbers themselves as well as individuals' attitude towards a figure. Moreover, one's previous experience and cognition determine the attitude to diseases. Schemas, being a set of ideas connected with cognitive structures employed in ordering, presenting, evaluating and using knowledge, are important in organizational settings since they help understand the behaviors of others, predict them and respond to them in an efficient way [47]. Since *the projection of image schemas onto abstract thought is mediated mainly by metaphor* and moreover *metaphor constitutes a crucial link between bodily experience and abstract reason* [48], in the following sections metaphors will be given a more detailed study.

4.4 Metaphors

Metaphors can be defined as *mappings from one conceptual domain to another* [49]. There are various ways of researching metaphors. As far as the metaphors presented in this chapter are concerned, the eco-linguistic theory of metaphor by Döring and Nerlich [50] is taken into account. It encompasses approaches from the Cognitive Theory of Metaphor, the Interaction Theory of Metaphor, Blumenberg's (1960) historical study of metaphor (Metaphorologie), and the Textual Theory of Metaphor. The intersection of all these approaches makes it possible to study metaphors in various ways. As Döring and Nerlich underline, especially the theories of Weinrich and Blumenberg and the concept of *image fields* are useful for the discussion on metaphors: *image fields are the product of experiential and synergetic processes between an organism and an environment, the outcome of an active and ongoing engagement within environments* [51]. In addition, the ecolinguistic point of view shows languages not as fixed structures, *but as open systems and repositories of accumulated social and cultural experience* [51]. In the case of this research, metaphors are dynamic phenomena that adjust to the needs of those who rely on them in some socio-natural and cultural contexts [51].

As far as the functionality of metaphors in discourse is concerned, they are useful in discussing novel or difficult concepts since they rely on symbols that are well-known and recognized by people [52–54]. Thus, metaphors help disseminate scientific and medical knowledge among non-specialists [55]. Moreover, since metaphors make complicated issues more comprehendible [56, 57] and limit the fear of change [58], they can be useful in the discussion on food risks. Moreover, *metaphors do not answer questions, they rather pose new questions* [59]. Consequently, they have ambiguity [60] and some mystery [61] in themselves, they offer various interpretations [62] and perspectives [63], and, thus, they shape the way people perceive the reality [64].

Individuals used myths, metaphors and rituals to communicate information related to risks even in the ancient times [65]. Taking into account the modern usage of figurative linguistic tools, risks can be perceived in a metaphorical way since when one mentions risks, it means that people should avoid something or at least be careful about something [66]. In this case, taking into account such features as metaphorical attractiveness for readers and their ability to draw attention [67, 68] more easily than other linguistic devices, metaphors prove to be an efficient tool in risk communication. One way is to look at risks through the perspective of *Pending Danger* (Damocles' Sword) that is characterized by the artificiality of risk source, visible catastrophic potentiality and the threat of randomness as far as victims are concerned. The next perspective is the one called *Slow Killers* (Pandora's Box) that entail artificial elements in food, water or air that have delayed effects on one's health; they are determined by information coverage and are easy to blame. The third one can be named *Cost-Benefit Ratio* (Athena's Scale) and it is strictly related to monetary gains and losses, characterized by an asymmetry between risks and gains, directed at a variance of distribution and dominated by probabilistic thinking. The fourth element, *Avocational Thrill* (Hercules' Image), stresses individual's control over the degree of risk and individual skills determining the comprehension of danger, involving voluntary activity and having non-catastrophic consequences [38].

There are certain areas that are often described through metaphors. One of them is GMO (Genetically modified food). For example, food biotechnology is presented as Frankenfood [26, 69] or by using the figure of Frankenstein directly, e.g. *Frankenstein sul piatto o panacea per sfamare i poveri del pianeta?*[1] [70]. The next scientific topic that relies on metaphors in communication is the discussion on food-borne diseases.

4.5 Metaphors and Food-Borne Diseases

Metaphors and visual imagery belong to the most important ways of conceptualizing illnesses and diseases [71]. As far as the role of using metaphors in discussing illnesses is concerned, there are two issues that are raised in scientific investigations. Some scientists claim that metaphors help people deal with difficult

---

[1] Frankenstein on the plate or the panacea to feed the poor of the planet (translated by the current author).

situations, including illnesses. For example, in the face of life-threatening disease, people often perceive them in terms of a battle, with an illness being an enemy that must be destroyed [72]. In consequence, an illness is viewed as something that can be overcome by patients, an adversary that can be defeated. In contrast, some state that such a metaphorical representation may lead to a wrong perception of real dangers since using military metaphors and perceiving illnesses in terms of heroic fight may lead to the false assumptions on their real hazard [73]. Thus, in the next part of this paper an attempt is made to discuss how metaphors determine the perception and cognition of risks related to food-borne diseases. The reasons for the selection of the topic are as follows. First of all, food-borne disease receive extensive media coverage. Secondly, metaphors do not only represent how jour-nalists view the reality, but they also create the reality and shape one's perception and reaction to information [74]. Thus, the role of mass media in communicating food-borne diseases is very important since the way one uses metaphors may lead to political, social, and economic consequences [75].

It should also be stated that the research on communicating food-borne disease is characterized by high dynamism; since one single risk may alter the perception of other risks [11], the representation of food-borne disease is determined by other health-related hazards. Moreover, the selection of metaphors depends on the stage of disease spread. During an outbreak, both a war metaphor and a journey metaphor are popular, whereas the second stage is described by a metaphor of control, supernatural force and a global network frame [76]. Since warfare belongs to one of the most popular domains used in metaphors [77, 78] and allows us to describe such phenomena as competitiveness, frightening for stakeholders, a metaphor of war constitutes one of the most popular ways of understanding diseases [79, 80]. A domain of war is also used in a discussion on food-borne diseases. For example, a foot-and-mouth disease is symbolized by applying the metaphors of fighting, journey or race [55]. Moreover, a metaphor of war may be used to stress simul-taneously diseases and those fighting with diseases, both attackers and victims. For example, governments may be presented as the ones attacking diseases and suc-cessful in combat [76].

A metaphor of machine is also used in medicine [81]. For example, a healthy person is pictured as a properly functioning machine, whereas a disease is portrayed through the motif of machine failures [80]. In addition, the most often used met-aphors to depict diseases stem from natural forces of air, earth, fire and water. Thus, viruses are often characterized as earthquakes, floods or storms [82]. Diseases can also be perceived by means of a road metaphor [83]. Additionally, a journey metaphor is also used in the discourse on health problems since a disease may be portrayed as a physical entity heading towards a goal [76], with infecting a person being the main aim. Disease may also be described by animal metaphors. For example, a virus is a greyhound reflects the conceptual metaphor of a race [82].

4.6 Metaphors of Food-Borne Diseases—Research

The aim of the research is to pay attention to various metaphors used in the discourse on food-borne diseases in Italy. Thus, the author has investigated the articles published in the online versions of Italian daily newspapers as well as at special portals and websites devoted to food handling. As far as the methodology is concerned, the Internet search engines of mentioned information services have been used. After the examination of over 30 articles on food-borne disease published in various online sources (see Appendix), several metaphorical domains have been identified that are popular in discussing food-borne diseases.

4.6.1 Disease Is a War

There are various reasons why a war metaphor is used in discussing food-borne diseases. One of them is the unpredicted and sudden appearance of an illness. The other function of war metaphors is to draw one's attention to the necessity of undertaking immediate and determined actions to overcome diseases.

In Francia viene segnalato un focolaio di diarrea emorragica che colpisce 16 persone

(In France the outbreak of the bloody diarrhea hit 16 people) Il Fatto Alimentare

Il motivo per cui solo nel maggio 2011 è scoppiata l'epidemia…

(The reason why the epidemics exploded in May 2011…) Il Fatto Alimentare

La salmonellosi colpisce in prevalenza i bambini

(Salmonella strikes mainly children) Il Fatto Alimentare

È è assai difficile che un'infezione sfugga e riesca a diffondersi

(It is very difficult since the infection escapes and succeeds in spreading) Il Fatto Alimentare

I polli e le uova alla diossina invadono l'Europa nel giugno del 1999

(The chickens and eggs with dioxin invaded Europe in June 1999) GRECO

Batterio killer, la Germania fa dietrofront

(Bacteria killer, Germany turns around) Corriere della Sera

Gli "agguati" dei microbi ai cibi

(The ambush of food microbes) Corriere della Sera

Quali sono i principali nemici da cui guardarsi?

(What are the main enemies that one should be beware of?) Corriere della Sera

Perché a due settimane dall'esplosione dell'epidemia di E.coli in mezza Europa, il panico ha ormai contagiato tutta l'economia

(Because 2 weeks after the explosion of *E. Coli* epidemics in half of Europe the panic has almost infected the whole economy) La Repubblica.

4.6.2 Disease Is a Sport

Since many prototypical sports, including soccer and rugby, have evolved from fighting [84], the domains of war and sport have many issues in common. They are both connected with gaining advantage, winning as well as being classified, etc. For example, the place in any classifications may be used to show the virulence of bacteria in food.

L'epidemia ha così guadagnato il secondo posto nella classifica delle intossicazioni alimentari europee dopo la Mucca pazza

(The epidemic has thus gained the second place after the mad cow disease in the classification of food poisonings) Il Fatto Alimentare.

4.6.3 Disease Is a Physical Entity

Food products often undergo personification in the discourse on food-borne disease. This perspective draws the attention of stakeholders to the source of infection. Their role in the infectious chain is strengthened by using trial metaphors to show their guilt or innocence in food poisoning.

In questi 36 giorni mentre cetrioli, pomodori e lattuga erano banditi dalle tavole, i cittadini hanno continuato a consumare germogli e ad ammalarsi

(During these 36 days when cucumbers, tomatoes and lettuce were banned from the tables people continued to eat sprouts and they became ill) Il Fatto Alimentare

I tedeschi "assolvono" i cetrioli spagnoli

(The Germans "discharge" the Spanish cucumbers) Corriere della Sera

I cetrioli importati dalla Spagna, inizialmente sospettati di aver provocato l'epidemia...

(The cucumbers imported from Spain, initially suspected of having caused the epidemics...) Corriere della Sera

Batterio killer, assolti i germogli di soia discharge

(Killer bacteria, the accused soy sprouts discharged) Il Messagero.

Apart from food products themselves, viruses are also personalized. They are portrayed in an anthropomorphic way to stress the fact that they constitute a part of individuals' life:

Un carico medio di lavatrice contiene 100 milioni di E. Coli oltre a Norovirus, Salmonella o Staphylococcus aureus. Per eradicare questi sgraditi ospiti della nostra lavanderia occorrono temperature di almeno 40 gradi, spesso in combinazione con adatti detergenti

(An average load of laundry contains 100 million of *E. Coli*, Norovirus, Salmonella and Staphylococcus aureus. To eliminate these unwelcome guests in our laundry, one should wash at the temperature of at least 40°, often together with suitable detergents) Corriere della Sera.

4.6.4 Disease Is a Journey

A journey metaphor is also used to discuss the role of change agents, especially the role of alternation and learning. Moreover, this metaphor encourages participants to join the activity [85]. In the case of food-borne diseases, it may serve the following functions. First of all, as far as food products are concerned, they are treated as vehicles, responsible for "transporting" diseases:

Pesci e formaggi molli e semimolli i principali "veicoli" alimentari

(Fish as well as soft and semisoft cheese are the main food vehicles) Il Fatto Alimentare

Il veicolo dell'infezione non è ancora stato identificato

(The vehicle of the infection has not been identified yet) Corriere della Sera
Secondly, placing a disease itself under scrutiny, it is pictured as a dynamic entity, able to travel very quickly and potentially infecting many people located in various places.

La contaminazione parte dal Belgio

(The contamination starts from Belgium) GRECO

Contemporaneamente si sospetta che il batterio abbia varcato l'oceano arrivando negli Stati uniti

(At the same time it is suspected that the bacteria has crossed the ocean and arrived in the USA) TGCOM24.

4.6.5 Disease is a Natural Disaster

Diseases may also be described by using the domain of natural disasters to show their power and vast consequences that are sometimes very difficult to estimate.

I "polli alla diossina", scoperti in Belgio nello scorso fine settimana, stanno provocando un terremoto politico

(The chickens with dioxin discovered in Belgium during the last weekend are causing a political earthquake) La Repubblica.

4.7 Discussion

Taking into account the investigated metaphors, it can be stated that they differ in the way they shape risk cognition. As has been presented in the empirical part of this investigation, different metaphors have diversified potential of drawing one's attention to various aspects of food-borne diseases. Taking into account the virulence of food-borne diseases, the domain of war is very powerful. Verbs such as *invade* or *attack* are used to show the strength and unpredictability of diseases. At the same time, such nouns as *ambush* and *turnabout* may denote simultaneously the "tactics" of diseases and those infected. Thus, a metaphor of war can also be used to picture the power of food authorities, doctors and patients in overcoming food poisonings. Taking into account the above-mentioned features of a war metaphor, it can be compared to a double-edged sword that can serve two functions in food risk communication; it can show the malevolent side of food-borne disease as well as the potential of human beings and their knowledge in fighting with these illnesses. To sum up, it should also be underlined that the mentioned war metaphors can be used not only to denote the features of diseases but also the determined attitude of victims. Thus, a war metaphor, depending on its use, can picture food-borne diseases as very risky and difficult to overcome, and, at the same time, as the ones that can be fought and defeated by individual or group strategies, if only the latter are eager to combat the virulence of bacteria in food. The same applies to the metaphorical domain of natural disasters and sports that can portray both diseases and the infected ones as having potential and strength. An important approach is to depict food-borne diseases as physical entities. This perspective, stressing e.g. the role of vegetables or raw meat in the process of food poisoning, may draw one's attention to the proper selection of products and the right application of adequate hygienic procedures in food preparation. Moreover, the cognitive impact of metaphors can be shaped by other metaphors as well. For example, the speed of bacteria spread can be highlighted by the use of a travel metaphor. This approach stresses a vast area of potential strike as well as high speed of germ dispersion and highlights the awareness of food poisoning in various geographical locations.

In consequence, it should be underlined that metaphors should be selected with great care by information writers since they determine individuals' perception of diseases, raising or lowering the riskiness of food-borne diseases in the eyes of the general public. Thus, the selection of metaphors should mirror the intentions of information providers. If e.g. health authorities want the general public to fight with food-borne disease, war metaphors should be selected. When the tempo of disease spread is to be highlighted, a metaphor of journey may serve this purpose. It should also be added that the chosen metaphors should be in line with individuals' cognition. Consequently, the domains used in metaphors should be recognized by the general public. As has already been discussed in this contribution, well-known domains facilitate the understanding of such complicated and multi-layered issues as food-borne diseases.

4.8 Summary

The aim of this chapter was to discuss the metaphorical dimension of food-borne diseases. The author concentrated on the examples coming from the Italian press that show the metaphoricity of the discourse on food risks. Taking into account the plurality of domains used in the creation of metaphorical information on disease, it can be stated that metaphors are a powerful tool in the discussion on food-borne diseases since the selection of metaphors determines the cognition of risks related to food consumption. Moreover, relying on well-known metaphors may determine one's attitude to risky situations and the subsequent actions related to disease treatment.

Articles Quoted in Providing Examples

Il Fatto Alimentare

http://www.ilfattoalimentare.it/storia-errori-epidemia-escherichia-coli-o104h4.html.
   http://www.ilfattoalimentare.it/salmonella-infezioni-cibi-efsa-sicirezza-alimentare.html.
   http://www.ilfattoalimentare.it/listeria-melone-vittime-usa-situazione-sotto-controllo.html.

GRECO

http://www.uniurb.it/giornalismo/lavori/greco/diossina.htm.

La Repubblica

http://www.repubblica.it/online/fatti/pollo/papitto/papitto.html.

Corriere Della Sera

http://www.corriere.it/salute/nutrizione/11_maggio_31/batterio-killer-spagna_9c5
eb4f2-8b6d-11e0-93d0-5db6d859c804.shtml.
http://www.corriere.it/esteri/10_giugno_26/farkas-salmone-ogm_744e9812-
8149-11df-9a47-00144f02aabe.shtml.
http://www.corriere.it/salute/nutrizione/11_giugno_17/infezioni-alimentari-
precauzioni-sparvoli_46878916-8dd7-11e0-b332-ace1587d6ad6.shtml.
http://www.corriere.it/salute/11_novembre_15/lavatrice-bassa-temperatura-
peccarisi_8ce5c046-0ae4-11e1-8371-eb51678ca784.shtml.
http://finanza.repubblica.it/News_Dettaglio.aspx?code=645&dt=2011-06-09&
src=TLB.
http://www.ilmessaggero.it/home_nelmondo/batterio_killer_assolti_i_germogli_
di_soia_fazio_controlli_a_tappeto_no_blocco_import/notizie/151808.shtml.

Appendix

Confini della sicurezza: http://www.uniurb.it/giornalismo/lavori/greco/titoli.htm.
Corriere della Sera: http://www.corriere.it/.
Il Fatto Alimentare: http://www.ilfattoalimentare.it/.
Il Messagero: http://www.ilmessaggero.it/.
Il Piccolo: http://ilpiccolo.gelocal.it/.
La Repubblica: http://www.repubblica.it/.
Panorama: http://www.panorama.it/.
TGCOM24: http://www.tgcom24.mediaset.it/.

References

1. Castells, M. (2004). *The network society. A cross-cultural perspective.* Cheltenham: Edward Elgar.
2. Bielenia-Grajewska, M. (2011). Corporate networking. In G. A. Barnett (Ed.), *Encyclopedia of social networks* (pp. 182–187). Thousand Oaks, CA: SAGE Publications Inc.
3. Gannon, M. J. (2008). *Paradoxes of culture and globalization.* Los Angeles: SAGE.
4. Bauman, Z. (2003). *Liquid modernity.* Cambridge: Polity Press.
5. Bielenia-Grajewska, M. (2010). The linguistic dimension of expatriatism—hybrid environment, hybrid linguistic identity. *European Journal of Cross-Cultural Competence and Management, 1*(2/3), 212–231.

6. Misra, S., & Stokols, D. (2012). Psychological and health outcomes of perceived information overload. *Environment and Behavior, 44*(6), 737–759.
7. Skyrme, D. (1999). *Knowledge networking: Creating the collaborative enterprise.* Oxford: Butterworth-Heinemann.
8. Strother, J. B., Ulijn, J. M., & Fazal, Z. (2012). *Information overload: An international challenge for professional engineers and technical communicators.* Hoboken, NJ: Wiley.
9. Van Heghe, H. (2006). *Learning to swim in information.* Penryn: Ecademy Press.
10. Shapiro, C., & Varian, H. R. (2006). *Information rules. Le regole dell' economia dell' informazione.* Milan: Etas.
11. Beck, U. (1986). *Risikogesellschaft. Auf dem Weg in eine andere Moderne.* Frankfurt am Main: Suhrkamp.
12. Bielenia-Grajewska, M. (2013b). Risk society. In K. B. Penuel, M. Statler & R. Hagen (Eds.), *Encyclopedia of crisis management.* Thousand Oaks, CA: SAGE Publications, Inc.
13. Toffler, A., & Toffler, H. (1995). *Budowa nowej cywilizacji. Polityka trzeciej fali.* Poznań: Zysk i S-ka Wydawnictwo s.c.
14. Sontag, S. (1989). *Illness as metaphor and AIDS and its metaphors.* New York: Picador.
15. Tepperwein, K. (2011). *Was dir deine Krankheit sagen will.* Munich: Mvg Verlag.
16. Rosenberg, Ch. E., & Golden, J. (1997). *Framing disease: Studies in cultural history.* New Brunswick: Rutgers University Press.
17. Tulloch, J., & Lupton, D. (2002). Consuming risk, consuming science: The case of GM foods. *Journal of Consumer Culture, 2*(3), 363–383.
18. Thobaben, M. (2010). Causes and prevention of foodborne illness. *Home Health Care Management & Practice, 22*(7), 533–535.
19. Balzaretti, C. M., & Marzano, M. A. (2013). Prevention of travel-related foodborne diseases: Microbiological risk assessment of food handlers and ready-to-eat foods in northern Italy airport restaurants. *Food Control, 29,* 202–207.
20. Gauci, Ch., & Gauci, A. A. (2005). What does the food handler in the home know about salmonellosis and food safety? *The Journal of the Royal Society for the Promotion of Health, 125*(3), 136–142.
21. Heath, R. L. (1994). *Management of corporate communication: From interpersonal contacts to external affairs.* Hillsdale, NJ: Lawrence Erlbaum Associates.
22. Leppin, A., & Aro, A. R. (2009). Risk perceptions related to SARS and avian influenza: Theoretical foundations of current empirical research. *International Journal of Behavioral Medicine, 16*(1), 7–29.
23. Morgan, C., & Younis, T. (1989). Policy analysis model of the problem of food-borne illness. *The Journal of the Royal Society for the Promotion of Health, 109*(5), 166–170.
24. Frewer, L. J., Howard, Ch., & Shepherd, R. (1996). Effective communication about genetic engineering and food. *British Food Journal, 98*(4/5), 48–52.
25. Gauthier, E. (2009). Social representations of risk in the food irradiation debate in Canada, 1986–2002. *Science Communication, 32*(3), 295–329.
26. Mcinerney, C., Bird, N., & Nucci, M. (2004). The flow of scientific knowledge from lab to the lay public: The case of genetically modified food. *Science Communication, 26*(1), 44–74.
27. Nucci, M. L., Cuite, C. L., & Hallman, W. K. (2009). When good food goes bad: Television network news and the spinach recall of 2006. *Science Communication, 31*(2), 238–265.
28. Jacob, C. J., Lok, C., Morley, K., & Powell, D. A. (2011). Government management of two media-facilitated crises involving dioxin contamination of food. *Public Understanding of Science, 20*(2), 261–269.
29. Maurer, J., Byrd-Bredbenner, C., & Grasso, D. (2007). "Ask before you eat"—development of an educational campaign on food allergies. *Social Marketing Quarterly, 13*(2), 48–70.
30. Bech-Larsen, T., & Aschemann-Witzel, J. (2012). A macromarketing perspective on food safety regulation: The Danish ban on trans-fatty acids. *Journal of Macromarketing, 32*(2), 208–219.
31. Harrari, P., & Legge, K. (2001). *Psychology and health.* Oxford: Heinemann Educational Publishers.

32. Pitts, M., & Phillips, K. (2001). *The psychology of health. An introduction.* Hove: Routledge.
33. Vartti, A. M., Oenema, A., Schreck, M., Uutela, A., de Zwart, O., Brug, J., & Aro, A. R. (2009). SARS knowledge, perceptions, and behaviors: A comparison between Finns and the Dutch during the SARS outbreak in 2003. *International Journal of Behavioral Medicine, 16* (1), 41–48.
34. Gray, L., MacDonald, C., Mackie, B., Paton, D., Johnston, D., & Baker, M. G. (2012). Community responses to communication campaigns for influenza A (H1N1): A focus group study. *BMC Public Health, 12*(205), 1–12.
35. Miles, S., Braxton, D. S., & Frewer, L. J. (1999). Public perceptions about microbiological hazards in food. *British Food Journal, 101*(10), 744–762.
36. Waring, A. E., & Glendon, I. (2006). *Managing risk: Critical issues for survival and success into the 21st century.* London: Thomson Learning.
37. De Zwart, O., Veldhuijzen, I. K., Elam, G., Aro, A. R., Abraham, T., Bishop, G. D., et al. (2009). Perceived threat, risk perception, and efficacy beliefs related to SARS and other (emerging) infectious diseases: Results of an international survey. *International Journal of Behavioral Medicine, 16*(1), 30–40.
38. Renn, O., & Rohrman, B. (2000). Risk perception research-an introduction. In O. Renn & B. Rohrman (Eds.), *Cross-cultural risk perception: A survey of empirical studies* (pp. 11–54). Dordrecht: Kluwer Academic Publishers.
39. Kühberger, A., & Tanner, C. (2010). Risky choice framing: Task versions and a comparison of prospect theory and fuzzy-trace theory. *Journal of Behavioral Decision Making, 23*, 314–329.
40. Renn, O. (2010). Risk communication: Insights and requirements for designing successful communication programs on health and environmental hazards. In R. L. Heath & H. D. O'Hair (Eds.), *Handbook of risk and crisis communication* (pp. 90–98). New York: Routledge.
41. Elledge, B. L., Brand, M., Regens, J. L., & Boatright, D. T. (2008). Implications of public understanding of avian influenza for fostering effective risk communication. *Health Promotion Practice, 9*(4), 54S–59S.
42. Wagner, W., & Kronberger, N. (2001). Killer Tomatoes! Collective symbolic coping with biotechnology. In K. Deaux & G. Philogène (Eds.), *Representations of the Social* (pp. 147–164). Oxford: Blackwell Publishers Ltd.
43. De Fiore, L. (2005). Comunicare il rischio in medicina, dialogare sul rischio per la salute. *Bolletino d'informazione sui farmaci, 3*, 126–129.
44. Powell, D. A., Jacob, C. J., & Chapman, B. J. (2011). Enhancing food safety culture to reduce rates of foodborne illness. *Food Control, 22*, 817–822.
45. Tiozzo, B., Mari, S., Magaudda, P., Arzenton, V., Capozza, D., Neresini, F., & Ravarotto, L. (2011). Development and evaluation of a risk-communication campaign on salmonellosis. *Food Control, 22*, 109–117.
46. Amberg, S. M., & Hall, T. E. (2010). Precision and rhetoric in media reporting about contamination in farmed salmon. *Science Communication, 32*(4), 489–513.
47. Bielenia-Grajewska, M. (2013a). Schemas theory. In E. H. Kessler (Ed.), *Encyclopedia of management theory.* Thousand Oaks, CA: SAGE Publications, Inc.
48. Boers, F. (1996). *Spatial repositions and metaphor: A cognitive semantic journey along the up-down and the front-back dimensions.* Tübingen: Gunter Narr Verlag.
49. Lakoff, G., & Turner, M. (1989). *More than cool reason. A field guide to poetic metaphor.* Chicago: University of Chicago Press.
50. Döring, M., & Nerlich, B. (2008). An outbreak of poetry: Mapping cultural responses to foot and mouth disease in the UK 2001. In A. Fill & H. Penz (Eds.), *Sustaining language: Essays in applied ecolinguistics* (pp. 181–200). Vienna: LIT Verlag.
51. Döring, M., & Nerlich, B. (2005). *Assessing the typology of semantic change: From linguistic fields to ecolinguistics* (pp. 55–68). VI: Logos and Language.
52. Bielenia-Grajewska, M. (2009). The role of metaphors in the language of investment banking. *Iberica, 17*, 139–156.
53. Brown, V. (1994). The economy as text. In R. Backhouse (Ed.), *New directions in economic methodology* (pp. 368–382). London: Routledge.

54. Mladenov, I. (2006). *Conceptualizing metaphors. On Charles Peirce's marginalia*. Abingdon: Routledge.
55. Larson, B. M. H., Nerlich, B., & Wallis, P. (2005). Metaphor and biorisks: The war on infectious diseases and invasive species. *Science Communication, 26*, 243–268.
56. Tilley, Ch Y. (1999). *Metaphor and material culture*. Malden, MA: Blackwell Publishers Ltd.
57. Wee, L. (2005). Constructing the source: Metaphor as a discourse strategy. *Discourse Studies, 7*(3), 363–384.
58. Zaltman, G., & Zaltman, L. (2008). *Marketing Metaphoria: What deep metaphors reveal about the minds of consumers*. Harvard, MA: Harvard Business School Publishing.
59. Etzold, V., & Buswick, T. (2008). Metaphors in strategy. *Business Strategy Series, 9*(5), 279–284.
60. Spicer, A., & Alvesson, M. (2011). Metaphors for leadership. In M. Alvesson & A. Spicer (Eds.), *Metaphors we lead by: Understanding leadership in the real world* (pp. 31–50). Abingdon: Routledge.
61. Black, M. (1993). More about metaphor. In A. Ortony (Ed.), *Metaphor and thought* (pp. 19–41). Cambridge: Cambridge University Press.
62. Schon, D. A. (1969). *Invention and the revolution of ideas*. London: Tavistock Publications Ltd.
63. Cacciari, C. (1998). Why do we speak metaphorically? Reflections on the functions of metaphor in discourse and reasoning. In A. N. Katz, C. Cacciari, R. W. Gibbs, & M. Turner (Eds.), *Figurative language and thought* (pp. 119–157). New York: Oxford University Press.
64. Cornelissen, J. P. (2008). Metaphor. In R. Thorpe & R. Holt (Eds.), *The SAGE dictionary of qualitative management research* (pp. 128–129). London: SAGE Publications Ltd.
65. Heath, R. L., & O'Hair, H. D. (2010). The significance of risk and crisis communication. In R. L. Heath & H. D. O'Hair (Eds.), *Handbook of risk and crisis communication* (pp. 5–30). New York: Routledge.
66. Thompson, P. B. (2012). Ethics and risk communication. *Science Communication, 34*(5), 618–641.
67. Charteris-Black, J. (2007). *The communication of leadership: The design of leadership style*. Abingdon: Routledge.
68. Denning, S. (2007). *The secret language of leadership: How leaders inspire action through narrative*. San Francisco: CA, John Willey and Sons.
69. Leiss, W. and Powell, D. (1997). *Mad cows and mother's milk: The perils of poor risk communication*. Kingston: McGill-Queens' University Press.
70. Farkas, A. (2010). Il "supersalmone" Ogm spacca gli Usa. Corriere della Sera. Retrieved June 5, 2012 from http://www.corriere.it/esteri/10_giugno_26/farkas-salmone-ogm_744e9812-8149-11df-9a47-00144f02aabe.shtml.
71. Lupton, D. (2003). *Medicine as culture*. London: SAGE Publications Ltd.
72. Kielhofner, G. (2008). *Model of human occupation: Theory and application*. Baltimore, MD: Lippincott Williams & Wilkins.
73. Bury, M. (2005). *Health and illness*. Cambridge: Polity Press.
74. Gravengaard, G. (2012). The metaphors journalists live by: Journalists' conceptualisation of newswork. *Journalism, 13*(8), 1064–1082.
75. Williams, A. E., Davidson, R., & Chivers Yochim, E. (2011). Who's to blame when a business fails? How journalistic death metaphors influence responsibility? *Journalism & Mass Communication Quarterly, 88*(3), 541–561.
76. Martin de la Rosa, V. (2008). The persuasive use of rhetorical devices in the reporting of 'Avian Flu'. *Vigo International Journal of Applied Linguistics, 5*, 87–105.
77. MacFarlane, B. (1999). Re-evaluating the realist conception of war as a business metaphor. *Teaching Business Ethics, 3*, 27–35.
78. Monin, N., & Monin, J. (1997). Rhetoric and action: When a literary drama tells the organization's story. *Journal of Organizational Change Management, 10*(1), 47–60.
79. Goschler, J. (2005). Embodiment and body metaphors. *Metaphorik.de, 09*, 33–52.

80. Nordgren, A. (2001). *Responsible genetics: The moral responsibility of geneticists for the consequences of human genetics research*. Dordrecht: Kluwer Academic Publishers.
81. Segal, J. Z. (2005). *Health and the rhetoric of medicine*. Carbondale: Southern Illinois University.
82. Nerlich, B., & Halliday, Ch. (2007). Avian flu: The creation of expectations in the interplay between science and the media. *Sociology of Health & Illness, 29*(1), 46–65.
83. Schiefer, M. (2006). *Die metaphorische Sprache in der Medizin: Metaphorische Konzeptualisierungen in der Medizin und ihre ethischen Implikationen untersucht anhand von Arztbriefanalysen*. Zürich: LIT Verlag.
84. Kövecses, Z. (2010). *Metaphor: A practical introduction*. New York: Oxford University Press.
85. Casell, C., & Lee, B. (2012). Driving, steering, leading, and defending: Journey and warfare. metaphors of change agency in trade union learning initiatives. *Journal of Applied Behavioral Science, 48*(2), 248–271.

Chapter 5
User-Centred Design as a Risk Management Tool

Damien J. Williams and Martin Groen

Abstract Numerous risks have been identified to and from the design process. The most prominent approach to address design shortcomings is user-centred design (UCD); however, current implementations of UCD tend to overemphasise differences between users which make it less acceptable as a viable approach to the process of design and the goal of risk management. Three cases are discussed that illustrate a progression of design approaches from a lack of consideration of the user (accommodation; Control room design) to a more limited consideration of user needs through the consultation of guidelines (e-commerce website design) to an attempt to consider user needs through detailed study (assimilation; Consumer product design). It is proposed that the assimilation of user task behaviour in the design life-cycle will ensure the inclusion of usability considerations, thereby enabling UCD to be utilised as a risk management tool.

5.1 Introduction

Humans utilize a myriad of artificial objects or artefacts that have been designed to assist in realising goals and satisfying needs. These artefacts range from large-scale environments (i.e. control rooms of energy plants) to small-scale, commercial products (i.e. "white goods" such as kitchen appliances and "brown goods" such as audio-visual entertainment equipment) and virtual products (i.e. online shopping websites). Consequently, design activity affects every aspect of our day-to-day lives [1]. Moreover, of central importance to design is the concept of usability [2, 3]. Usability is defined as "the extent to which a product can be used by specified users

D.J. Williams (✉)
School of Medicine, University of St Andrews, St Andrews, Scotland
e-mail: djw11@st-andrews.ac.uk

M. Groen
Utrecht Institute of Linguistics OTS, Utrecht University, Utrecht, Netherlands
e-mail: m.g.m.groen@uu.nl

© Springer-Verlag Berlin Heidelberg 2015
J.-M. Mercantini and C. Faucher (eds.), *Risk and Cognition*,
Intelligent Systems Reference Library 80, DOI 10.1007/978-3-662-45704-7_5

to achieve specified goals with effectiveness, efficiency and satisfaction in a specified context of use" [4, p. 2]. From a design perspective, Noyes [2] noted that all
artefacts designed for human use should meet the user's basic requirements in terms
of usability, including:

1. *Ease* with which the artefact can be used
2. *Effectiveness* in allowing the user to achieve their goals
3. *Likability* which refers to whether or not the artefact is pleasing to use

These three factors highlight the objective (1 and 2) and subjective (3) components of usability. Thus, the rationale behind the concept of usability and the goal
of design is to "enhance human abilities, to support human limitations and to meet
the subjective, affective component that is unique to humans" [2, p. 63]. Moreover,
Altom [5] noted that the general risk associated with ignoring usability might be
that users will not use the product.

The focus on designing with those people that will actually use an artefact is
called *user-centred design* (UCD) [6]. A failure to implement UCD could potentially expose users and all interested and affected parties to a variety of risks. For
instance, at an individual level, considerations of usability can, most importantly,
reduce the risk of physical harm to users [7, 9] and prevent errors in using the
product [10]. In addition, the implementation of UCD can improve user satisfaction,
which is seen as increasingly important in design (pleasure-based approach) [11,
12]. For example, when a system offered relevant services that matched the needs
and expectations of the users, customer satisfaction was increased by 40 % [13].

Moreover, a consideration of usability can have a significant impact at the
organisational level. For instance, a variety of sources have expressed dissatisfaction with the design of artefacts, leading to a low adoption or usage of these
artefacts as reflected in the following accounts:

1. It has been estimated that several billions of dollars are lost in both public and
   private-sector organisations as a result of not considering user needs and
   requirements [14]
2. In terms of the international competitive market, Wakeford [15] noted that
   British organisations are at an increased risk of losing out to US competitors,
   because the former do not implement approaches that enable the design of
   artefacts that meet users' needs and requirements
3. Poor design of the built environment costs the UK millions of pounds every
   year [16]
4. Two-thirds of Americans lose interest in technology products because they are
   too complex to set up and operate. And only 13 % of the public believe that
   technology products are easy to use [17]
5. E-retailers lose approximately $4 billion in revenue due to poor web site usability
   meaning that prospective customers do not complete their transaction [18]
6. Only between 1 % [19] and 5 % [20] of new products can be considered
   successful

7. Poor computer interface design significantly contributed to the 1999 UK Passport Agency crisis which reportedly cost £12.6 million, including: staff overtime, compensation, and umbrellas for people who waited outside passport offices in the rain. While the supplier of the computer system agreed to pay £2.45 million of the costs, the remainder was covered by the taxpayer [21]

In order to design artefacts that successfully achieve their goal of supporting users it is crucial that potential risks are identified and dealt with at the earliest possible stage of the design process. A substantial risk to and from the design process may arise as a result of not taking appropriate steps to identify and implement users' needs and requirements into the final design. Thus, UCD can be considered a risk (and cost) management tool [5].

The current chapter will set out the rationale for the adoption of UCD as a risk management tool. To begin, a consideration of the concept of risk will identify the role that design can play. This will highlight the association between risk and design, which will then be explored further to identify the utility of UCD in the design process and its potential role in managing many of the risks to and from the design process. A brief critique of current implementations of UCD will provide justification for the implementation of an alternative perspective in which a consideration of the users' task behaviour is a necessary requirement. Finally, three cases are discussed in which interested and affected parties are exposed to various risks, and how a consideration of user-centred aspects that has improved/could improve the effectiveness of the design, thereby assisting the establishment of task-objectives and reducing exposure to risk.

5.2 The Association Between Risk and Design

The environment within which any human endeavour is undertaken is inherently risky [22], this includes the design endeavour itself [23]. Indeed, risk is a fundamental consideration in "a vast range of decision making situations, from allocating wealth to safeguarding public health, from waging war to planning a family, from paying insurance premiums to wearing a seatbelt, from planting corn to marketing cornflakes" [24, p. 2]. Consequently, an understanding of risk is believed to enable the planning of actions despite not knowing about the future [25]; however, this is easy to say and harder to do due to the historical and ongoing debate regarding *what is risk?*

In general, risk can be understood either as a statistical value or a synonym for danger or threat [26] which reflects the sharply contrasting understanding held by experts (i.e. risk assessors) and non-experts (members of the public), respectively [27]. Indeed, the traditional approach to the conceptualization of risk is to define risk as a two-dimensional value (probability and consequence) [28–31] known as the probabilistic approach. The belief is that these values are objective, analytic and rational, whereas anyone who does not subscribe to this approach upholds an erroneous "risk perception" that is subjective, hypothetical, emotional, and irrational [32].

The main argument against objective risk is that those individuals affected by risks often think differently and have different descriptions of the situation than experts [33]. Moreover, there is a substantial literature questioning the suitability of the concept of objective risk in risk management [1, 34–37] mainly highlighting the importance of the value laden nature of subjective risk [38] adopted by non-experts [27]. For instance, Thompson [37] noted that "a reasonable person's concept of risk ... is better suited to ... risk management than are probabilistic concepts ... any suggestion that probabilistic concepts of risk should become the basis of risk management ... is a regressive pursuit of false Gods".

This perspective is supported by the large body of research from the psychometric paradigm which has found that the concept of subjective risk can be understood in terms of four superordinate dimensions: "dread risk", unknown risks', unnatural and immoral risk, and the number of people exposed to a given risk (for details see [39–45]). Despite the apparent limitations identified with this body of research (see [46] for a discussion) it has highlighted the need to account for subjective risk when considering any risk issue including risk management and design [27, 47]. Indeed, within the consumer product literature, it has been identified that in order to design a safe product it is necessary to acknowledge that it is the subjective perceptions of an individual that will influence their future actions, rather than any real consideration of the objective risk [48, 49].

From the viewpoint of risk management, it is therefore necessary to understand what risk means to those individuals who will be "managed". By and large, this group of individuals will not be experts holding an "objective" understanding of risk, but will have a "subjective" understanding of risk. However, the material available from the psychometric paradigm regarding subjective risk is not amenable to risk management efforts when designing artefacts, as it cannot be easily implemented in the design process. Nonetheless, Williams and Noyes [47] referred to the work by Heimrich Kanis and colleagues [7, 8, 50–52] (discussed in more detail in the third case study on consumer product design) which illustrates how an understanding of user interaction with an artefact [50] can identify the (subjective) risk(s) associated with the artefact (usage-centred design [50]).

The underlying rationale for a usage-centred approach is that usability problems can evoke unsafe situations [8]. While this was in reference to consumer products, it is equally applicable to many other artefacts. What is more, this can be extrapolated to: usability problems can invoke risky situations. This implies that artefacts that are less useable could not only lead to physical harm to the user (as is the focus of the work by Kanis and colleagues) but, as highlighted in the definition of risk earlier, increase the potential for harm to profit, psychological well-being, status, reputation, et cetera. While not all potential risky usage scenarios can be prevented [53] as it is not possible to anticipate how the user will actually use the artefact, the most effective way to overcome this problem is to ensure that the design approach assumes a stance that focuses on assimilating user actions, needs, and requirements, which may include a number of potential risky usage scenarios. Indeed, Howarth [49] suggests that the most effective safety measures are those that operate directly on behaviour. Thus, an approach that focuses on the activities of the user of an artefact and considers the perception and evaluation of risk during artefact usage [7] presents a viable tool in risk management efforts as an understanding of user activities is a useful aid in the reduction of risk [27] and the design of useable artefacts. Moreover, this approach can be utilized during an iterative design process [54] using models or prototypes [52].

5.3 The Role of Users in the Design Process

The traditional approach to design has been to place emphasis on designing arte-facts that meet functional requirements [55] with the user merely considered as another resource that has to be optimized in order to meet the goals of operation [2]. Norman [56] highlighted the limitations of such an approach through a multitude of examples of design errors, which he referred to as the problem of pathological design. While these examples were rather innocuous (e.g. the design of the fascia of a door to indicate how to open and close it) merely posing a risk to user satisfaction/ frustration, pathological design could, dependent on the context, pose a greater risk at the individual and organisational level. Busby and Hibberd [57, p. 137] noted that a failure to take account of "... the natural characteristics and behaviours of users is failing (design) in its most basic purpose". Furthermore, van Duijne et al. [58, p. 246] emphasised that in risk management "the focus should not only be on the product properties (hazards), but also on the interaction of a person with the product". These viewpoints are consistent with UCD wherein the user's roles and responsibilities are viewed as fundamental to ensuring design success and are given priority in the design process [2] with the effect of reducing the risk of designing an unusable artefact.

Considering the number of years that have passed since the original concerns about traditional, functional design have been expressed, one would expect that the practice of UCD would be fully integrated into routine design practice. Unfortu-nately, this is not the case. It is argued that this results from an ill-conceived approach to UCD, which is elaborated in the following sections.

5.4 UCD as a Risk Management Tool

Up to this point, risk has mainly been addressed in terms of the potential for risk resulting from the use of an artefact; however, the origins of this risk are deeply rooted in the design process itself [23]. For instance, it is important that correct design decisions are made as early as possible in the design process [59]. An early, poor choice that is later decided to be revised, poses a risk at the organisational level in terms of adding further expenditure that will not be recovered, leading to extra time which will ultimately delay product release [60].

Indeed, there is growing realisation of the importance of UCD in mitigating such risks [14]. For instance, Skelton and Thamhain [14] distinguished between two classes of risk in the design project. First, "class I risks" are either directly or indirectly the result of complexities in the design of an artefact that lead to complexities in the design process (e.g. high cognitive load,[1] low agreement on design objectives, time pressure, peer group pressure, poor design specification or degree of usability) that impact on the success of the overall design process. Secondly, "class II risks" constitute a set of issues that may arise as a consequence of the realisation of class I risks (e.g. schedule slippages and budget overruns). Indeed, Skelton and Thamhain's field observation indicated that project managers often fail to anticipate class I risks, rather, they deal with problems after they have occurred, in the form of class II risks. This is akin to the 'band aid effect' described by Williams and Noyes [1] in which design activities are reactive (reacting to a class II risk) rather than proactive (proactively addressing potential type I risks). It was concluded that employing a UCD approach would enable designers to anticipate class I risks thereby facilitating the design of products that improve user-product interaction.

User involvement is typically utilised when problems arise during the latter stages of design. However, this will offer "at best limited corrective actions" [62, p. 299]. Moreover, such an approach would add considerably to the overall cost in terms of resource (financial, time, etc.) compared to the initial investment associated with a UCD approach, thereby posing a risk to business profit. In addition, potentially risky usage scenarios will only be addressed after a number of users have run into problems caused by using the artefact leading to various kinds of risks on the part of the producer and the user of the artefact. This illustrates the importance of integrating UCD throughout product development, particularly in the earliest stages where risk and uncertainty (regarding the nature of the product and its market success) are especially high [63] and where user involvement, through the exploration and definition of user needs and requirements [64] is most beneficial [65].

5.5 Involving the User in Design

The key concept in UCD is that the design of the artefact should assimilate the needs, wants, and limitations of the (intended) user. Katz-Haas [66, pp. 12–13] defined UCD as a "philosophy and a process [...] that places the person (as opposed to the 'thing') at the centre; it is a process that focuses on cognitive factors (such as perception, memory, learning, problem-solving, etc.) as they come into play during peoples' interactions with things". The problem of this rather open-ended definition and other similar definitions (e.g. [67]) is that it is difficult to determine which of

---

[1] Cognitive load refers to the demands on working memory storage and processing of information when a person works on a task [61].

these cognitive factors need to be considered in the design of an artefact. If one succeeds in selecting the appropriate cognitive factors, the next problem becomes the determination of the extent of sufficient consideration of the cognitive factor to inform the design of the artefact adequately. That is, how much of the cognitive factor needs to be understood in order to decide whether this understanding is sufficient for current purposes?

In this respect it is important to reiterate the design goal of the artefact: it should enable the user (or users in the case of artefacts aimed at collaboration, such as video conferencing) of that artefact to realise some goal that needs to be achieved. It is important to have a clear understanding of these goals, as without them it would be near impossible to find out *a priori* whether a specific design of an artefact actually enables a user to realise their goal(s). There are some proposals that seem to forego this important consideration and instead focus on epistemological issues (i.e. ecological approaches e.g. [68, 69]), of humans operating in an environment in which they find themselves when they are conducting tasks. According to this conception, objectivity needs to be avoided as it precludes an understanding of the "subjectivity of an agent as a prerequisite for his construction of the objective world in a communicative interaction with it" [68, p. 16]. This, however, leads to a very relativistic approach to the study of design, which even calls into question the whole enterprise of it. That is, if you cannot infer certain design requirements from the observation of humans conducting a task, which is what these ecological approaches seem to argue, then why bother designing products for large groups of people. In other words, specifying design requirements necessitates objectifying some aspects of the task praxis in order to inform designers what they should include in the product or to facilitate its use.

Despite the pragmatic advantages of a UCD approach, the philosophy is not universally accepted. For instance, Phibes [70] noted that such an approach merely leads to systems being designed to fulfil the whimsical needs of the individual. Consequently, he indicated that the UCD concept is frequently used in such a way as to be devoid of meaning, and would be better discarded. The argument here appears not to be against the principles of UCD, but the way in which they are implemented. That is, UCD seems to be conceptualised as a call for the design to include individual differences of users in the design of the artefact. In a review of qualitative, quantitative, and field studies Kujala [71] concluded that the effects of user involvement in the design process seem to be positive, but complicated. Plowman et al. [72] reported similar positive effects of user involvement in the computer supported cooperative work field.

Alternatively, Norman [73, p. 45] identified activity-centred design as a more practical approach to design arguing that "good behavioral organization reflects human activity structure, not dictionary classification". If the effective execution of a task is taken as the point of reference for including user data, then there is a natural limitation in the extent of different psychological phenomena that need to be considered in each case. The UCD question then becomes: which of the observed *task-based* behaviours are necessary to include in the design of the artefact. Consequently, the design goal of the artefact is then to assimilate the task-activities of the user [74], which is congruent with the usage-centred approach to designing for risk [27].

As previously highlighted, it is necessary for the user to be rooted firmly at the centre of the design process to prevent *ad hoc* design and the associated accumulation of cost. While user involvement is expensive [5], the longer-term benefits far-outweigh the initial expense. For instance, although the figures are somewhat dated Mayhew and Bias [75] illustrate the extent of the financial risk that for every dollar spent to resolve a problem during the early stages of product design, $10 would have to be spent on the same problem during the latter stages of product development. Moreover, they noted that if the same problem has to be resolved after release the costs rise to $100 or more. One way to help prevent the accumulation of costs through inappropriate design decisions being made and subsequently rectified at later stages is through the implementation of UCD principles from the beginning of the design lifecycle.

Many attempts to include user requirements in the design process have been ill-conceived. For instance, one way in which a consideration of user needs and requirements has been implemented, particularly in the field of human-machine interfaces, is for the designer to apply their own assumptions and stereotypes about how people use systems [76]. However, designers are not typical users [56] and cannot, therefore, rely on or expect to have an intimate knowledge of the "typical" user [77]. Moreover, Nemire [78, p. 7] suggested that "to prevent inaccurate mental models[2] from leading users into hazardous situations…designers should follow product safety guidelines to eliminate or reduce the hazards". While adherence to safety guidelines may reduce hazards/risks, reliance on guidelines (of any form) will not itself eliminate those risks. Indeed, Kontogiannis and Embrey [80] stated that it is not sufficient to design *for* users, it is also necessary to design *with* users. Thus, users should be a central component of the design process: "the more user involvement, the less chance that the final design will fail its users" [5, p. 16]. However, unlike the recommendations from participatory design in which users are members of the design team [81], the suggestion here is that the user be involved through user testing and evaluation. One of the potential limitations associated with the sole reliance on the participatory design approach is that the user representatives may become biased by those around them, leading to an expectation bias when they are involved in the testing and evaluation.

---

[2] According to Norman [79] people formulate mental models when interacting with an artefact, which then provides predictive and explanatory power for understanding the interaction. He further notes that while these models are constrained by the users' knowledge, experience, an information processing capabilities, and therefore not necessarily technically accurate, they are generally functional and are continually modified throughout the interaction.

5.6 Who Are the Users?

UCD prescribes a process that centres on the user of the artefact but, who are they? Within the literature there are many approaches to design that take into consideration, for example, customer requirements (see [82]) in order to identify customer-specific design requirements (customer-focused design). However, it is often the case that these approaches do not consider the needs and requirements of the user of the product, but rather corporate (customer) requirements that come from internal sources such as marketing, finance, manufacturing, and service, that may then influence the form or function of an artefact [82]. What is more, end-user requirements are seen merely as expectations that the user has of the product. It is essential that the conceptualisation of user expectations include relevant aspects of usability, functionality, reliability, safety, efficiency, and requisite user competencies [62, 64]. Gershenson and Stauffer [82, p. 103] correctly asserted that: "it is necessary to consider all customers, not just end-users, from the beginning of the design process [...] this practice saves time and money in the end, since it reduces downstream changes". Thus, it is not suggested that the design process be handed over to the users rather that all interested and affected parties be involved from the earliest stages, with a consideration of the user firmly rooted at the centre of the design activity [83].

An important consideration is that a wide range of possible user groups are involved in the design process to ensure that designs are usable by people irrespective of age or ability. This philosophy is consistent with the aims of inclusive design [84]. This does not imply that it is necessary to design to meet the needs of every conceivable user rather that it is recommend that a consideration of the various needs and requirements of potential user groups would be beneficial both for design and the user. Thus, the aims of inclusive design are congruent with, and can therefore be comfortably embedded within, and the practice of UCD.

The benefits of inclusive design are evident in the statistics that indicate that two often neglected user groups, "older people" (50+, see [85]) and disabled people, represent a considerable combined spending power (£330 billion, see [86]). In the future, the ratio of older people compared to younger people is expected to increase considerably worldwide from 1:12 in 1950 to 1:4 in 2050, as reported by the United Nations Population Division. Older people and disabled people often battle with a range of impairments that could influence the use of current artefacts. For instance, in the UK only 3 % of disabled people are wheelchair-bound, however 8.7 million have some degree of hearing loss, and one million have learning difficulties [86]. Additionally, one of the potential consequences of aging is the onset of a variety of impairments including the reduction in auditory and visual capacities, strength, co-ordination, and manual dexterity. Yelding [87] asserted that it is not these impairments that are the problem, but the poor design of artefacts that transform them into a source of disability. Consequently, in working towards the successful implementation of UCD, it would be necessary to incorporate inclusive design principles in order to appropriately design artefacts that are usable by all potential user groups. Further, when considering older and disabled users their active participation is even more crucial as designers frequently overlook their particular needs [88].

Not designing for these often overlooked potential users, would risk missing out on a substantial sector of the market. In particular, as the older user group represent a growing market segment [89] the design of artefacts must cater for their needs and requirements. While such an approach would impose an extra cost, it would be relatively small and would soon be off-set by the creation of a wider market. Accordingly, by understanding the real needs of potential user groups through the effective implementation of UCD, better artefacts can be designed that will benefit the user, designer, and companies and industries. Not only are monetary goals served (e.g. companies focusing on design are more profitable see [90]), but also a larger potential group of users is served by providing a more usable artefact.

5.7 Implementation of UCD

In the following section three case studies will be reviewed that describe the effect of choices in design practice. This work is evaluated with a focus on the aspects of assimilation or accommodation. That is: is the user expected to accommodate to the artefact, or is the artefact designed to assimilate the task behaviour of the user?

5.7.1 Case Study 1: Control Room Design

The purpose of this case study is to provide a brief overview of control room design issues identified through the investigation of a major incident at the Three Mile Island nuclear power plant.

At about 4 am March 28 1979 a major accident occurred at the Three Mile Island Unit 2 (TMI-2) nuclear power plant in Pennsylvania (see [91] for a detailed account). The problem arose as a result of the malfunction of one of the automatic relief valves, which resulted in an interruption of the flow of coolant water to the main pumps that subsequently damaged the reactor, as it was supposed to open, relieve the pressure, and then close automatically [92]. Although the TMI-2 plant suffered a severe meltdown (the most dangerous kind of nuclear power accident) there was no loss of life, and only a small amount of radioactive material was released; however, the cost to the operating companies and insurers was estimated to be in the region of 1 billion dollars [2].

Subsequent analysis of TMI-2 (see [93]) indicated that it was caused by a multitude of factors. While operator error was reported to have influenced the way events were managed, component failures and grossly inadequate control room design were also causal factors in the escalation of the incident [21, 93]. Harris [94] noted that most control rooms were designed with little attention to the needs of the operators. Indeed, Noyes [2, 95] identified a number of failures in human-machine interaction from the TMI-2 accident, these included:

1. *Information overload*. Operators were bombarded with information from a large number of flashing displays and a cacophony of undifferentiated auditory alarms, with no means of suppressing the unimportant ones (see [96])
2. *Control panel design*. Controls and instruments were organised in an inconvenient and illogical manner, such that the operators had to scan approximately 1,600 windows and gauges
3. *Indicators*. Several of the displays had gone "off scale" indicating the seriousness of the situation, but provided no information regarding the extent of the problem
4. *Printer failure*. Useful data about the situation was lost due to printer failure, and when the printer was restored, it was running more than 2 hours behind events

The analysis shows that users were forced to accommodate their behaviour to the inappropriate operating specifications of the control room, which resulted in a heavy (cognitive) burden being placed on them.

The importance of control room design is highlighted by the statistics resulting from accident investigations indicating that between 20 and 50 % of all accidents and incidents can be attributed to design failures; however, Wilpert [62] believed that the actual percentage is likely to be much higher. More specifically, the analysis of the TMI-2 incident identifies human performance as a critical component in control room safety [93]. Stone et al. [21, p. 10] stated that "The incident [TMI-2] could have been prevented if the control panels had been designed to provide the operators with the necessary information to enable them to perform their tasks efficiently and correctly".

This case study suggests that in order to ensure control rooms are operable and reduce the risk of serious incidents occurring in the future, their design should incorporate an understanding of the performance and information needs of operators, by focusing on task-based activities. This would ensure the provision of on-display content and format (including functional grouping, and colour coding) "off-normal" indicator provision (using colour displays) and accessibility of instruments [94] to support serious error and incident recovery [21] when necessary.

5.7.2 Case Study 2: Retail Website Design

The success of an online organisation, especially if it is not well-known, is mainly dependent on the design of its website [97–99]. Developing an appropriate website is akin to building the customer interface for the entire organisation [100–102]. Consequently, organisations cannot assume that just because they have a site customers will use it, as users have become less patient with unusable websites and are aware of alternative possibilities [103]. The site is therefore a chief defining factor in terms of their willingness to make a purchase [97]. Thus, an understanding of web-based interaction/activity is necessary in order to design the site in a way that it is straightforward for customers to perform the necessary tasks [101]. This will ultimately increase customer satisfaction, retention, and loyalty [104] and reduce the risk to the economic viability of the organisation [103].

When using a site, users need to find their way around the site and locate necessary information. Thus, central to the design of an effective and usable site are the ease of navigation and search. Indeed, among the website design mistakes identified by Nielsen [105] are the issues of poor navigability and confusing content. These issues will impact on the user as they will require users to remember information found at each step, thereby increasing their cognitive load [106]. Consequently, sites that are designed to reduce the cognitive load on the user will be more effective as they will enable users to perform the necessary tasks with less effort.

An abundance of heuristic recommendations are available for website design (e.g. [104]). Gehrke and Turban [97] conducted a literature survey and identified five categories: page-loading speed, business content, navigation efficiency, security, and marketing/customer focus. Next, a user survey was conducted to identify the importance of the categories from the perspective of the user. The results of the two surveys were congruent. It was concluded that users want content and service to be provided fast, rather than in a technologically advanced or artistic manner. However, relying solely on these recommendations and similar rules-of-thumb is not advised, when one considers the large number of usability guidelines for web design that are currently available [104]. Not only do the guidelines contradict each other, their abundance makes it difficult to determine which are the most important.

Studying the tasks users are required to undertake generates appropriate design briefs. As users uncover problems, designers should correct these problems and then continue with further testing [107]. Miles et al. [108] view a site as a type of (purchasing) decision support system, they note that it is necessary to consider all stages of the purchasing process. Nielsen [104] described a study in which users were required to perform various tasks on 20 large and small e-commerce sites. A total of 496 attempts were made with only 56 % being successful. Such findings suggest that e-commerce organisations may be losing a large number of potential sales simply because their sites do not effectively support the user at all stages of the purchasing process.

In conclusion, a site should assimilate the users' tasks and their understanding of the offered products and services, thereby designing for usability (see [2]) and an optimal user experience [104]. It is acknowledged that a website may be visited by a range of users, all with different needs [109] which will undoubtedly make the implementation of a UCD approach difficult. Nonetheless, failing to design the website for, and adapted to, the needs and requirements of the user will increase the risk of frustrating users, which could result in the loss of (potential) customers and, in the long run, may even risk the profitability of the organisation [103, 104].

This case study illustrates that studying user behaviour whilst interacting with the website could provide insight into how to prevent unintended effects owing to the design of the site. Assimilating this behavioural data could prevent the need for users to accommodate their behaviour to the site when purchasing products or services. As the data shows this could lead to missed opportunities and potential losses for the company involved.

5.7.3 Case Study 3: Consumer Product Design

Safety is a fundamentally important issue in all areas of design. Within the area of consumer products there is now a great expectation among consumers with regard to the safety of the diverse range of products they encounter on a daily basis [52]. Consequently, designers and manufacturers of consumer products go to considerable lengths to produce safe products. However, given the importance of consumer perceptions of a product to its market success, there is also a need to adopt the users' safety needs and demands as the driving force in the endeavour to create safer products.

Historically, the matter of including user requirements has largely occurred through the consideration of human characteristics as found in handbooks. Kanis [50] discussed a number of studies involving a variety of consumer products, which identified the lack of, or at best a loose association between human characteristics and user activities. It was concluded that general user characteristics, as sourced from handbooks, serve little purpose in the prediction of user activities. Empirical data regarding the actual usage of a product is imperative (see [110] for a discussion of empirical methods).

In addition to the use of handbooks, in-depth interviews with victims of consumer product accidents (involving kitchen utensils, do-it-yourself products and personal care items) and accident reconstructions can be used to study the adverse outcomes associated with product use (see [111]). In so doing, a number of interesting issues were identified including, for example, product features that did not adequately reflect the product's actual mode of operation, which led users to believe that they were safe to use. However, these results are based on *ad hoc* reconstructions of the episodes of use, not on observation of actual usage. Furthermore, it was identified that only those people involved in accidents with consumer products participated in the study, which limits an understanding of how the product is used by users in general. A subsequent approach by van Duijne et al. [112] was to investigate the actual use of a number of common do-it-yourself products (hand saw, a chisel and hammer, an electric drill and a screwdriver). Semi-structured interviews were used in order to better understand user actions and the risks users perceived. It was suggested that safety was not necessarily the primary motivator in the use of the tools, rather it was the products' 'ease of use', that is whether the product design facilitates easy operation.

The design of safe consumer products is dependent on knowledge about actual user activities [112]. While the approaches adopted above attempt a user-centred approach, the way in which they have been implemented limits the applicability of the outcomes. The deployment of user trials as a design tool is concerned with the identification of the variability in usage [113]. However, in order to create an accurate picture of potential usage it would be necessary to involve different types of users who might come into contact with the product (i.e. experts and novices, the elderly, the young, physically and mentally impaired, etc.) consistent with the principles of inclusive design. Thus, through a UCD approach the knowledge gained from investigating user-product interaction can reveal subtle ways in which product characteristics are perceived by users. What is more, this process can be adequately undertaken during design using observational studies with related products, models, or prototypes without the need to study the actual occurrence of accidents. As a result, it would be possible to provide designers with insights into the requirements necessary to assimilate the variation in user activities in the design in order to prevent risks due to unintended usage.

This case study shows that a more in-depth understanding of user behaviour whilst using the product could provide insight into how to minimise risk and prevent unintended usage by assimilating the understanding gained from the users into the design. Relying on the user to accommodate their behaviour to the usage of these devices could have played a salient role in the occurrence of unsafe behaviour.

5.7.4 From Accommodation to Assimilation

The three case study's outlined here illustrate a progression of design approaches from a lack of consideration of the user (accommodation; Case study 1) to a more limited consideration of user needs through the sole consultation of pre-existing heuristics/guidelines (Case study 2) to an attempt to consider user needs through detailed empirical study (assimilation; Case study 3). This final approach illustrates how the concept of UCD has been appropriately incorporated in the design of the product to support users to realise their tasks goals. What is hopefully demonstrated here is the need to assimilate user task behaviour when designing, so as to ensure not only the safety of primary and secondary users, but also the credibility of designers and manufacturers.

5.8 Recommendations for Design

User involvement has generally be found to have positive effects in design, particularly on user satisfaction, and evidence suggests that taking users as a primary source of information is an effective means of capturing design requirements [114]. Moreover, the adoption of UCD can help mitigate risks that inherently accompany artefact design; this is a considerable advantage of UCD. Indeed, presenting usability as risk management has been identified as an effective way of engaging business people (see [5]); however, the common implementation of UCD has led to an overemphasis on individual differences (e.g. [67]). It is, therefore, understandable that organisations are not keen to accept UCD in that incarnation, as it is too costly for artefacts to be designed for individuals. Also, due to this focus on individual differences, the task requirements for which the artefact was supposed to be designed are often overlooked resulting in ineffectual artefacts.

Our view of the most appropriate approach to ensure sufficient inclusion of usability considerations in the design process, thereby enabling UCD to be utilised as a risk management tool, are encompassed in Nielsen's [115] recommendations:

1. Consider the larger context
2. Know the user

   - Individual user characteristics
   - The user's current task
   - Functional analysis
   - Evolution of the user

3. Competitive analysis
4. Setting usability goals
5. Participatory design
6. Coordinated design of the total artefact

   - Standards
   - Product identity

7. Guidelines and heuristic analysis
8. Prototyping
9. Empirical testing
10. Iterative design

    - Capture the design rationale

11. Collect feedback from field use.

The approach outlined in this chapter would sit within "Step 2: Know the user" whereby a task-oriented [73] or usage-centred design [50] approach would not only facilitate the understanding of user behaviour in a given task/with a given artefact, but also highlight risk(s) associated with an artefact. This knowledge of the user, along with their continual involvement (i.e. through empirical user testing) can then be implemented at subsequent stages (3–11).

5.9 Conclusion

The low usability and acceptance of artefacts due to inappropriate design presents considerable long-term risks to the users as well as the organisations and institutions that produce them. Designing artefacts that expect consumers to accommodate their task behaviour to the artefact implies that the designers expect users to be able to recognise and respond to the expected activities included in the design. As can be seen in many examples in day-to-day practice, and in the three case study's presented here, this often fails resulting in the emergence of risks and leading to what is referred to as human error (see [53]), and the subsequent low acceptance and use of new or adapted products and services.

The crucial issue is that users cannot be expected to be able to infer all the intended functionality of an artefact: they are in specific task contexts, will most likely not know as much about the artefact as the designer, and will have different needs and requirements when using the artefact. However, comprehending the intended functionality of an artefact is necessary in order to effectively, efficiently, and safely use it when realising tasks, and thus design must be undertaken *with* users.

Despite the apparent difficulties of implementing UCD, it does not mean that we should give up on its adoption in the design process. Rather, a different conceptualisation is presented in the current chapter that recommends that designers study user behaviour in order to provide appropriate data regarding artefact usage that can then be subsequently utilised in the design process. Thus, to manage risk to and from the design process, and design successful (in terms of being usable and sought after in the marketplace) artefacts requires a thorough understanding of the relevant task behaviours of the user(s) which can then be assimilated in the design of the artefact. Indeed, the substantial human factors toolkit (see [116]) means that we are well-equipped to address these challenges and implement UCD as a risk management tool leading to safer and more effective products and services. Specifically, the utilisation of carefully designed controlled experiments could provide the required understanding needed to design safe, effective, efficient, and likeable artefacts.

References

1. Williams, D. J., & Noyes, J. M. (2007). Effect of risk perception on decision-making: Implications for the provision of risk information. *Theoretical Issues in Ergonomics Science, 8*(1), 1–35.
2. Noyes, J. M. (2001). *Designing for humans*. Hove, UK: Psychology Press.
3. Wiklund, M. E. (Ed.). (1994). *Usability in practice: How companies develop user-friendly product*. San Diego: Academic Press.
4. International Organization for Standardization. (1998). ISO 9241. Ergonomic requirements for office work and visual display terminals—Part 11: Guidance on Usability. International Organization for Standardization. Retrieved May 20, 2007, from http://www.iso.org/iso/catalogue_detail.htm?csnumber=16883
5. Altom, T. (2007). Usability as risk management. *Interactions, 14*(2), 16–17.
6. Norman, D. A., & Draper, S. W. (Eds.). (1986). *User centered system design: New perspectives on human-computer interaction*. Hillsdale, NJ: Lawrence Erlbaum Associates.
7. van Duijne, F. H., Kanis, H., Hale, A. R., & Green, B. S. (2008). Risk perception in the usage of electrically powered gardening tools. *Safety Science, 46*(1), 104–118.
8. van Duijne, F. H., Hale, A., Kanis, H., & Green, B. (2007). Design for safety: Involving users' perspectives. Redesign proposals for gas lamps using a pierceable cartridge. *Safety Science, 45*(1–2), 253–281.
9. Gagg, C. (2005). Domestic product failures: Case studies. *Engineering Failure Analysis, 12* (5), 784–807.

10. Baber, C., & Stanton, N. A. (2002). Task analysis for error identification: Theory, method and validation. *Theoretical Issues in Ergonomics Science, 3*(2), 212–227.
11. Jordan, P. W. (1998). *An introduction to usability.* London: CRC Press.
12. Holt, J., & Lock, S. (2008). Understanding and deconstructing pleasure: A hierarchical approach. Paper presented at the Computer Human Interaction 2008 conference. Retrieved September 10, 2011, from www.chi2008.org/altchisystem/submissions/submission_jane66_0.pdf
13. Usability.net. (2006). The business case for usability. Usability.net. Retrieved March 8, 2007, from http://www.usabilitynet.org/management/c_business.htm
14. Skelton, T. M., & Thamhain, H. J. (2005). User-centered design as a risk management tool in new technology product development. In *Proceedings of the IEEE International Engineering Management Conference* (vol. 2, pp. 690–694). doi: 10.1109/IEMC.2005.1559237
15. Wakeford, N. (2005). People-centered innovations. *Strategic Direction, 21*(4), 30–32.
16. Sorrell, J., Simmons, R., Desyllas, J., & Nicholson, R. (2006). *The cost of bad design.* London: Commission for Architecture and the Built Environment.
17. Philips Index (2004). Calibrating the convergence of healthcare, lifestyle and technology. Philips Index. Retrieved March 8, 2007, from http://www.philipsindex.ca/
18. Bringula, R. P., & Basa, R. S. (2011). Factors affecting faculty web portal usability. *Educational Technology & Society, 14*(4), 253–265.
19. den Buurman, R. (1997). User-centred design of smart products. *Ergonomics, 40*(10), 1159–1169.
20. Lyall, S. (2006). Browser (website design). *The Architectural Review, 220*(1317), 95–95.
21. Stone, D., Jarrett, C., Woodroffe, M., & Minocha, S. (2005). *User interface design and evaluation.* San Francisco, CA: Elsevier.
22. Hillson, D., & Murray-Webster, R. (2006). *Understanding and managing risk attitude.* Aldershot, UK: Gower Publishing Limited.
23. Jerrard, B., & Barnes, N. (2006). Risk in design: Key issues in the design literature. *The Design Journal, 9*(2), 25–38.
24. Bernstein, P. L. (1998). *Against the gods: The remarkable story of risk.* New York, NY: Wiley.
25. Steele, J. (2004). *Risks and legal theory.* Oxford, UK: Hart Publishing.
26. Oppe, S. (1988). The concept of risk: A decision theoretic approach. *Ergonomics, 31*(4), 435–440.
27. Williams, D. J. (2006). Conceptualization of risk. In W. Karwowski (Ed.), *International encyclopedia of ergonomics and human factors* (2nd ed., Vol. 1, pp. 301–303). London: CRC Press.
28. Starr, C. (1969). Social benefit versus technological risk. *Science, 165*(3899), 1232–1238.
29. Barki, H., Rivard, S., & Talbot, J. (1993). Toward an assessment of software development risk. *Journal of Management Information Systems, 10*(2), 203–225.
30. Hoegberg, L. (1998). Risk perception, safety goals and regulatory decision-making. *Reliability Engineering and System Safety, 59*(1), 135–139.
31. Naoe, K. (2008). Design culture and acceptable risk. In P. E. Vermaas, P. Kroes, A. Light, & S. A. Moore (Eds.), *Philosophy and design: From engineering to architecture* (pp. 119–130). London: Springer Verlag.
32. Slovic, P. (1997). Trust, emotion, sex, politics and science. In M. H. Bazerman, D. M. Messick, A. E. Tenbrunsel, & K. A. Wade-Benzoni (Eds.), *Environment, ethics and behaviour* (pp. 277–313). San Francisco, CA: Lexington Press.
33. Rehmann-Sutter, C. (1998). Involving others: Towards an ethical concept of risk. *Risk: Health, Safety & Environment, 9.* Retrieved June 24, 2006, from http://heinonline.org/HOL/LandingPage?collection=journals&handle=hein.journals/risk9&div=16&id=&page=
34. Hansson, S. O. (2005). Seven myths of risk. *Risk Management: An International Journal, 7* (2), 7–17.

35. Shrader-Frechette, K. S. (1990). Perceived risks versus actual risks: Managing hazards through negotiation. *Risk: Health, Safety & Environment, 1*. Retrieved June 24, 2006, from http://heinonline.org/HOL/LandingPage?collection=journals&handle=hein.journals/risk1&div=34&id=&page=

36. Valverde, L. J. (1991). The cognitive status of risk: A response to Thompson. *Risk: Issues in Health & Safety, 2*. Retrieved June 24, 2006, from http://heinonline.org/HOL/LandingPage?collection=journals&handle=hein.journals/risk2&div=29&id=&page=

37. Thompson, P. B. (1990). Risk objectivism and risk subjectivism: When are risks real? *Risk: Issues in Health & Safety, 1*. Retrieved June 24, 2006, from http://heinonline.org/HOL/LandingPage?collection=journals&handle=hein.journals/risk1&div=9&id=&page=

38. Cross, F. B. (1992). The risk of reliance on perceived risk. *Risk: Issues in Health & Safety, 3*. Retrieved June 24, 2006, from http://heinonline.org/HOL/LandingPage?collection=journals&handle=hein.journals/risk3&div=11&id=&page=

39. Fischhoff, B., Slovic, P., Lichtenstein, S., Read, S., & Coombs, B. (1978). How safe is safe enough? A psychometric study of attitudes toward technological risks and benefits. *Policy Sciences, 9*(2), 127–152.

40. Morgan, M. G., Slovic, P., Nair, I., Geisler, D., MacGregor, D., Fischhoff, B., et al. (1985). Powerline frequency electric and magnetic fields: A pilot study of risk perception. *Risk Analysis, 5*(2), 139–149.

41. Kraus, P. P., & Slovic, P. (1988). Taxonomic analysis of perceived risk: Modeling individual and group perceptions within homogeneous hazard domains. *Risk Analysis, 8*(3), 435–455.

42. Mullett, E., Duquesnoy, C., Raiff, P., Fahrasmane, R., & Namur, E. (1993). The evaluative factor of risk perception. *Journal of Applied Social Psychology, 23*(19), 1594–1605.

43. Sjöberg, L., & Torell, G. (1993). The development of risk acceptance and moral valuation. *Scandinavian Journal of Psychology, 34*(3), 223–236.

44. Slovic, P. (1987). Perception of risk. *Science, 236*(4799), 280–285.

45. Sjöberg, L. (2000). Factors in risk perception. *Risk Analysis, 20*(1), 1–11.

46. Millstein, S. G., & Halpern-Felsher, B. L. (2002). Perceptions of risk and vulnerability. *Journal of Adolescent Health, 31*(1), 10–27.

47. Williams, D. J., & Noyes, J. M. (2011). Reducing the risk to consumers: Implications for designing safe consumer products. In N. A. Stanton, W. Karwowski, & M. Soares (Eds.), *Handbook of human factors in consumer product design: Uses and applications* (vol. 1, pp. 3–21). Boca Raton: CRC Press.

48. Mitchell, V.-W. (1999). Consumer perceived risk: Conceptualisations and models. *European Journal of Marketing, 33*(1/2), 163–195.

49. Howarth, C. I. (1988). The relationship between objective risk, subjective risk and behaviour. *Ergonomics, 31*(4), 527–535.

50. Kanis, H. (1998). Usage centred research for everyday product design. *Applied Ergonomics, 29*(1), 75–82.

51. Kanis, H. (2002). Can design-supportive research be scientific? *Ergonomics, 45*(14), 1037–1041.

52. van Duijne, F. H., Kanis, H., & Green, B. (2002). Risks in product use: Observations compared to accident statistics. *Injury Control and Safety Promotion, 9*(3), 185–191.

53. Reason, J. T. (1990). *Human error*. Cambridge, UK: Cambridge University Press.

54. McRoberts, S. (2005). Risk management of product safety. In *Proceedings of the 2005 IEEE Symposium on Product Safety Engineering* (pp. 65–71). doi: 10.1109/PSES.2005.1529524

55. Bevan, N. (1999). Quality in use: Meeting user needs for quality. *Journal of Systems and Software, 49*(1), 89–96.

56. Norman, D. A. (1988). *The psychology of everyday things*. Cambridge, MA: MIT Press.

57. Busby, J. S., & Hibberd, R. E. (2002). Mutual misconceptions between designers and operators of hazardous systems. *Research in Engineering Design, 13*(3), 32–138.

58. van Duijne, F. H., van Aken, D., & Schouten, E. G. (2008). Considerations in developing complete and quantified methods for risk assessment. *Safety Science, 46*(2), 245–254.

59. Chen, L., & Li, S. (2000). Modeling concurrent product design: A multifunctional team approach. *Concurrent Engineering-Research and Applications, 8*(3), 183–198.
60. Ullman, D. G. (2001). Robust decision making for engineering design. *Journal of Engineering Design, 12*(1), 3–13.
61. Schnotz, W., & Kürschner, C. (2007). A reconsideration of cognitive load theory. *Educational Psychology Review, 19*(4), 469–508.
62. Wilpert, B. (2007). Psychology and design processes. *Safety Science, 45*(1–2), 293–303.
63. Murphy, J. (2000). Assuring performance in E-commerce systems. In *Proceedings of the IEE 16th UK Telegraphic Symposium* (pp. 30/1–30/6). Retrieved September 11, 2011, www. eeng.dcu.ie/ ~ murphyj/publ/c14.pdf
64. Damodaran, L. (2001). Human factors in the digital world enhancing life style: The challenges for emerging technologies International. *Journal of Human-Computer Studies, 55* (4), 377–403.
65. Noyes, J. M., Starr, A. F., & Frankish, C. R. (1996). User involvement in the early stages of the development of an aircraft warning system. *Behaviour & Information Technology, 15*(2), 67–75.
66. Katz-Haas, R. (1998). Ten guidelines for user-centered web design. *Usability Interface, 5*(1), 12–13.
67. Black, A. (2006). The basics of user-centred design. Design Council. Retrieved February 13, 2007, from http://www.designcouncil.org.uk/en/About-Design/DesignTechniques/User-centred-design/
68. Muller, M., Shami, N. S., Millen, D., & Feinberg, J. (2010). We are all lurkers: Consuming behaviors among authors and readers in an enterprise file-sharing service. In Proceedings of the 16th ACM International Conference on Supporting Group Work, (pp. 201–210). New York: ACM Press. doi: 10.1145/1880071.1880106
69. Pluempavarn, P., Panteli, N., Joinson, A., Eubanks, D., Watts, L., & Dove, J. (2011). Social roles in online communities: Relations and trajectories. Paper presented at the 6th Mediterranean Conference on Information Systems, Nicosia, Cyprus. Retrieved October 4, 2012, from http://aisel.aisnet.org/mcis2011/47
70. Phibes, T. A. D. (2002). The perverse horrors of user-centric design. *Expert Systems, 19*(5), 295–298.
71. Kujala, S. (2008). Effective user involvement in product development by improving the analysis of user needs. *Behaviour & Information Technology, 27*(6), 457–473.
72. Plowman, L., Rogers, Y., & Ramage, M. (1995). What are workplace studies for? In H. Marmolm, Y. Sundblad, & K. Schmidt (Eds.), *Proceedings of the Fourth European Conference on Computer-Supported Cooperative Work* (pp. 309–324). Norwell: Kluwer Academic Publishers. Retrieved February 02, 2011, from http://www.ecscw.org/1995/20.pdf
73. Norman, D. A. (2006). Logic versus usage: The case for activity-centered design. *Interactions, 13*(6), P. 45 & 63.
74. Groen, M., & Noyes, J. (2011). Product design: User-centred versus a task-based approach. In W. Karwowski, M. Soares, & N. Stanton (Eds.), *Handbook of human factors in consumer product design* (Vol. 1, pp. 405–413). Boca Raton, FL: CRC Press.
75. Mayhew, D. J., & Bias, R. G. (1994). *Cost-justifying usability*. Boston, MA: Academic Press.
76. Darses, F., & Wolff, M. (2006). How do designers represent to themselves the users' needs? *Applied Ergonomics, 37*(6), 757–764.
77. Storer, I., & McDonagh, D. (2002). Embracing user-centred design: The real experience. In P. T. McCabe (Ed.), *Contemporary ergonomics* (pp. 307–314). London: Taylor and Francis.
78. Nemire, K. (2008). Roller coasters, mental models, and product safety. *Ergonomics in Design, 16*(4), 7–10.
79. Norman, D. A. (1983). Some observations on mental models. In D. Gentner & A. L. Stevens (Eds.), *Mental models* (pp. 7–14). Hillsdale, NJ: Lawrence Erlbaum Associates.
80. Kontogiannis, T., & Embrey, D. (1997). A user-centred design approach for introducing computer-based process information systems. *Applied Ergonomics, 28*(2), 109–119.

81. Zaphiris, P., Laghos, A., & Zacharia, G. (2009). Distributed construction through participatory design. In M. Khosrow-Pour (Ed.), *Encyclopedia of information science and technology* (2nd ed., pp. 1181–1185). Hershey, PA: Information Science Reference.
82. Gershenson, J. K., & Stauffer, L. A. (1999). A taxonomy for design requirements from corporate customers. *Research in Engineering Design, 11*(2), 103–115.
83. Williams, D. J. (2007). Risk and decision making. In M. J. Cook, J. M. Noyes, & Y. Masakowski (Eds.), *Decision making in complex systems* (pp. 43–54). Aldershot, UK: Ashgate.
84. Keates, S., & Clarkson, J. (2003). Design exclusion. In P. J. Clarkson, R. Coleman, S. Keates, & C. Lebbon (Eds.), *Inclusive design: Designing for the whole population* (pp. 88–107). London: Springer-Verlag.
85. Huppert, F. A. (2003). Designing for older users. In P. J. Clarkson, R. Coleman, S. Keates, & C. Lebbon (Eds.), *Inclusive design: Design for the whole population* (pp. 30–49). London, UK: Springer Verlag.
86. La Ferla, B., Hosking, I., & Sinclair, K. (2006). Whose design is it anyway? *Engineering Management, 16*(2), 10–13.
87. Yelding, D. (2003). Power to the people. In P. J. Clarkson, R. Coleman, S. Keates, & C. Lebbon (Eds.), *Inclusive design: Designing for the whole population* (pp. 108–119). London: Springer-Verlag.
88. Abascal, J., & Nicolle, C. (2001). Why inclusive design guidelines? In C. Nicolle & J. Abascal (Eds.), *Inclusive design for HCI* (pp. 3–13). London: CRC Press.
89. Newell, A. (2003). Inclusive design or assistive technology. In P. J. Clarkson, R. Coleman, S. Keates, & C. Lebbon (Eds.), *Inclusive design: Designing for the whole population* (pp. 172–181). London: Springer-Verlag.
90. Hertenstein, J. H., Platt, M. B., & Veryzer, R. W. (2005). The impact of industrial design effectiveness on corporate financial performance. *Journal of Product Innovation Management, 22*(1), 3–21.
91. Samuel, W. J. (2004). *Three Mile Island: A nuclear crisis in historical perspective*. Berkeley, CA: University of California Press.
92. Noyes, J. M., & Stanton, N. A. (1997). Engineering psychology: Contribution to system safety. *Computing & Control Engineering Journal, 8*(3), 107–112.
93. Malone, T. B., Kirkpatrick, M., Mallory, K., Eike, D., Johnson, J. H., & Walker, R. W. (1980). *Human factors evaluations of control room design and operator performance at Three Mile Island-2: Final report*. Ann Arbor, MI: University of Michigan Library.
94. Harris, D. H. (1984). Human factors success stories. In Proceedings of the 28th Annual Meeting of the Human Factors Society (pp. 1–5). Santa Monica: Human Factors Society. Retrieved May 6, 2005, from www.hfes.org/Web/PubPages/Harris.pdf
95. Noyes, J. M. (2001). Human error. In J. M. Noyes & M. L. Bransby (Eds.), *People in control: Human factors of control room operations* (pp. 3–15). London: IEE.
96. Noyes, J. M. (1998). Managing errors. In *Proceedings of the UKACC International Conference on Control '98* (vol. 1, pp. 578–583). doi: 10.1049/cp:19980293
97. Gehrke, D., & Turban, E. (1999). Determinants of successful website design: Relative importance and recommendations for effectiveness. In Proceedings of the 32nd Hawaii International Conference on System Sciences. doi: 10.1109/HICSS.1999.772943
98. Feindt, S., Jeffcoate, J., & Chappell, C. (2002). Identifying success factors for rapid growth in SME e-commerce. *Small Business Economics, 19*(1), 51–62.
99. Sebora, T., Lee, S., & Sukasame, N. (2009). Critical success factors for e-commerce entrepreneurship: An empirical study of Thailand. *Small Business Economics, 32*(3), 303–316.
100. Korper, S., & Ellis, J. (2001). *The e-commerce book: Building the e-empire* (2nd ed.) London: Academic Press.
101. Kamoun, F., & Halaweh, M. (2012). User interface design and E-commerce security perception: An empirical study. *International Journal of E-Business Research, 8*(2), 15–32.

102. Kuo, H. M., & Chen, C. W. (2011). Application of quality function deployment to improve the quality of internet shopping website interface design. *International Journal of Innovation in Computer and Information Control, 7*(1), 253–268.
103. Nah, F. F., & Davis, S. (2002). HCI research issues in E-commerce. *Journal of Electronic Commerce Research, 3*(3), 98–113.
104. Nielsen, J. (1999). *Designing web usability: The practice of simplicity.* Berkeley, CA: Peachpit Press.
105. Nielsen, J. (2005). The top ten web design mistakes of 2005. Jokob Nielsen's Alertbox. Retrieved March 8, 2007, from http://www.useit.com/alertbox/designmistakes.html
106. Conklin, J. (1987). Hypertext: An introduction and survey. *Computer, 20*(9), 17–41.
107. Corry, M. D., Frick, T. W., & Hansen, L. (1997). User-centered design and usability testing of a web site: An illustrative case study. *Educational Technology Research and Development, 45*(4), 65–76.
108. Miles, G. E., Howes, A., & Davies, A. (2000). A framework for understanding human factors in web-based electronic commerce. *International Journal of Human-Computer Studies, 52*(1), 131–163.
109. de Troyer, O. M. F., & Leune, C. J. (1998). WSDM: A User centered design method for web sites. *Computer Networks and ISDN Systems, 30*(1–7), 85–94.
110. Noyes, J. M., & Garland, K. J. (2006). Empirical methods: Experiments. In W. Karwowski (Ed.), *International encyclopaedia of ergonomics and human factors* (2nd ed., pp. 3119–3121). London: Taylor & Francis.
111. Weegels, M. F., & Kanis, H. (2000). Risk perception in consumer product use. *Accident Analysis and Prevention, 32*(3), 365–370.
112. van Duijne, F. H., Green, W. S., & Kanis, H. (2001). Risk perception: Let the user speak. In M. A. Hanson (Ed.), *Contemporary ergonomics 2001* (pp. 297–302). London: Taylor and Francis.
113. Kanis, H., & Vermeeren, A. P. O. S. (1996). Teaching user involved design in the Delft Curriculum. In S. A. Robertson (Ed.), *Contemporary ergonomics 1996* (pp. 98–103). London: Taylor and Francis.
114. Kujala, S. (2003). User involvement: A review of the benefits and challenges. *Behaviour & Information Technology, 22*(1), 1–16.
115. Nielsen, J. (1992). The usability engineering life cycle. *Computer, 25*(3), 12–22.
116. Noyes, J. M. (2004). The human factors toolkit. In C. Sandom & R. S. Harvey (Eds.), *Human factors for engineers* (pp. 57–79). London: IEE.

Chapter 6
Analysis of the User Behaviour When Interacting with Systems During Critical Situations

Yuska P.C. Aguiar, Maria de Fátima Q. Vieira, Edith Galy-Marie and Charles Santoni

Abstract Human error studies tend to focus on identifying the relationship between the human activity, its errors and consequences. Accidents and incidents report analysis has been the path followed by several authors in the human error studies field, as it will be discussed in this chapter. However, reports tend to detail technical aspect of the error occurrence but fail to explore the human-behaviour component that might have influenced it. In order to investigate the human behaviour and its relation with accidents and human errors the authors propose to observe individuals working during critical situations. This observation must adopt a methodological approach, and the authors advise to support it by an experimental protocol, to ensure a rigorous systematization of the data gathering and analysis of the human behaviour. This chapter presents a cognitive model conceived to support this approach, by investigating an individual's: characteristics; functional state; situation perception, decision-making approach and performance during task completion which accounts for the knowledge of the work situation. It also briefly presents its supporting experimental protocol, and discusses its application in the context of a decision making aid system, employed during maritime pollution crisis management.

Y.P.C. Aguiar
Department of Exact Sciences, UFPB Rio Tinto, Paraiba, Brazil
e-mail: yuska@dce.ufpb.br

M. de Fátima Q. Vieira (✉)
LIHM, Electrical Engineering Department, UFCG,
Campina Grande, Paraiba, Brazil
e-mail: fatima@dee.ufcg.edu.br

E. Galy-Marie
Département de Psychologie Cognitive et Expérimentale,
Centre de Recherche PsyCLE,
Aix-Marseille Université, Aix-En-Provence, France
e-mail: edith.marie@univ-provence.fr

C. Santoni
Laboratoire des Sciences de l'Information et des Systèmes,
Aix-Marseille Université, Avenue Escadrille Normandie-Niemen,
Marseille Cedex 20, France
e-mail: charles.santoni@polytech.univ-mrs.fr

© Springer-Verlag Berlin Heidelberg 2015   129
J.-M. Mercantini and C. Faucher (eds.), *Risk and Cognition*,
Intelligent Systems Reference Library 80, DOI 10.1007/978-3-662-45704-7_6

6.1 Introduction

It is widely accepted in the literature that the human error results from failures in the cognitive system. Those failures frequently happen during knowledge acquisition, which comprises a set of cognitive processes and activities that act on sensory information in order to interpret, classify and organize it. To prevent the error occurrence it is important to investigate the risk factors involved and understand the full context in which the error occurs, in order to identify which combined factors can influence the human behaviour and performance, and consequently the task outcome. Once being aware of the risk factors which can trigger the human error, and having identified the factors which may influence task performance, preventive measures can be put into place to: improve the process adopted when performing the task; the tools employed in doing it; and the human operator skills.

This research investigates the adoption of cognitive models to help understanding the cognitive process and anticipate risk-related behaviour in the work context. It focuses in analyzing the error context (when an operator of an industrial automated system is faced with unexpected work situations), combined with the analysis of typical time pressures, anxiety and altered emotional behaviour, in order to anticipate error occurrences. It follows a brief review of the literature which presents the human error as seen from different perspectives by different authors. These do not treat directly the proposed research problem, but give a foundation for the study.

Hollnagel [1] defines the error as the consequence of a faulty action which leads into unexpected results. According to Rasmussen [2], when a system presents an unsatisfactory response to a human action, which is different from the expected one, a human error has occurred. Yet, for Reason [3, 4], the human error occurs when the consequences of a fault cannot be assigned to external agents, exposing to risks: people, equipment and the surrounding environment. The authors adopt a combination of those views and assume that a cognitive model is at the basis of understanding the human interaction and the task outcome. The following cognitive models adopt complimentary approaches and share the notion that managing cognitive resources to perform a task and deciding which action to perform are influenced by multiple variables.

According to Norman and Draper [5], the human cognitive process happens in seven stages: formulating objective; formulating intension; specifying action; performing action; perceiving the status of the environment and evaluating results. Therefore, an individual can assess whether an achieved result corresponds to the intended one realizing when there was an error.

The Interaction model proposed by Norman and Baucom [6] defines an interaction as consisting of two phases: execution and evaluation. The execution phase translates objectives into intentions to perform a task of interest. This intention results into a sequence of actions (mental specifications) which are converted into actions in the physical world. The evaluation phase starts with the perception of the environment in response to the action performed. This perception is then interpreted and compared to the initial objectives.

The SRK (Skill-Rule-Knowledge) model, proposed by Rasmussen [2], describes qualitatively different modes of information processing during task performance, accounting for a behaviour based on skills; rules and knowledge.

The cognitive model proposed by Endsley [7] considers that task performance is determined by a decision made after the adequate understanding of the current situation. This decision is processed in three levels; element perception, element understanding and anticipation mechanisms, which allow to foresee the following state of the situation, given that the intended action is executed. Endsley model of mental activities follows a sequence that begins with detecting a sensory signal, followed by accessing the memory, performing logic and intuitive thinking, arriving at the decision making phase and completed by performing the action. System and task characteristics, such as complexity levels, as well as the risk involved in doing the task determine the work load and the levels of vigilance and cognitive control, thus determining the available cognitive resources or mental abilities [8]. Limiting the availability of such cognitive resources, through high spending elsewhere in the task, results in operator behaviour automation [9]; which in turn can lower performance during task execution.

Although the influence of these factors on the cognitive processes, and consequently on the operator performance is widely accepted, there is no explicit representation of those variables in the cognitive models found in the literature, which would help anticipate the human behaviour which leads into error. Given the need for a deeper understanding of the human behaviour, especially when working with risk situations, this chapter proposes a model which accounts for the above variables in an attempt to help understand the human behaviour in such work environments, and thus anticipate the error prone human behaviour. This model's main objective is to identify internal and external elements which affect operator behaviour and performance, allowing to measure the level of influence of each variable on the observed behaviour; thus enabling to propose changes in the working environment, particularly in the ergonomic aspect of the human interface component of the employed tools. The numbers of variables to account for in the study, as well as the analysis of the cross influence between those, are the main challenges of this research. Nonetheless an initial simplified model has been proposed and an experiment has been performed to test its impact on the human error study. Both the model and experiment are presented in this chapter.

6.1.1 Critical Systems Usability and Human Error Prevention

In the domain of critical systems, accident and incident report analysis is at the basis of the human error study. Although this is cited in the literature as the main approach taken in the field, as mentioned in the works of: Rasmussen [10], van Eekhout and Rouse [11] and Johnson and Rouse [12], according to Rasmussen [10], analyzing the human behaviour when faced with adversities during task performance is also an important source of information on the cognitive mechanisms and strategies employed during activity.

The authors of this chapter adopt the following definition for a critical system. This is any system whose failure could threaten human lives, the system's environment or the existence of the organisation which operates the system. Failure in this context means any potentially threatening system behaviour other than failure to conform to a specification. Examples of typical critical systems are command and control systems such as air-traffic control and electricity supply systems, disaster management systems among others.

On the other hand, critical situations are not exclusive to critical systems. In this work context, critical situations arise when decisions must be made and actions must be taken under time pressures and cognitive resources' constrains; and the failure to do so might also threaten human lives, the system's environment, or the existence of the organisation which operates the system.

Aiming to support the human error study by widening the range of information available beyond that obtained through report analysis, this chapter proposes the adoption of a method and protocol to support the observation and analysis of the human behaviour during the interaction in critical situations. It is proposed that an individual taking part in such experiment must be immersed into a work context that reproduces the conditions described in accident and incident reports. That is, an individual with a similar profile to the original human operator must be lead into performing the same task under similar conditions, whilst being observed according to the proposed protocol. This approach in observing the human behaviour when performing tasks in a critical situation is synthesized in the experimental protocol introduced later in this chapter. This protocol systematizes the observation planning, execution and documentation of the observation experiment.

From the usability point of view the investigation inherits the techniques employed in observing the interaction between systems and users when performing a predefined task under controlled conditions, i.e., during a usability test. During these tests, data is gathered to produce objective and subjective metrics, as part of a diagnostic about the system in use. In this research, cognitive psychology, supports the investigation of the relationship between the functional and emotional state of an operator in charge of a system, his (her) workload, and how the combination of these two influences: the error occurrence; the task outcome and the operator's performance. This investigation relies on the adequate choice of tools, available to work-psychology studies, in order to support data gathering and analysis, relevant to the understanding of the human behaviour.

6.2 Understanding the Human Behaviour

To understand the human behaviour, that is, the reactions of an individual during a critical situation, it is necessary to analyze this situation's characteristics and the individual's functional state in order to identify the relevant human behaviour and its components. These reactions happen in several observable levels such as: performance, emotional, and physiological. The tools employed in detecting and measuring these reactions can be organized similarly into one of three categories, according to the targeted information: subjective, performance and psycho-physiological measuring tools. This section presents the tools employed during the research.

6.2.1 Subjective Measures

Subjective measures enable estimating parameters which are related to an individual's feelings and reactions induced by a situation (or context), such as: emotions, functional state, and workload perception.

Subjective Workload Measures There are two most commonly used techniques for measuring the subjective mental workload. The first one, NASA-Task Load Index [13], classes the workload into six subscales: Mental Demand, Physical Demand, Temporal Demand, Own Performance, Effort, and Frustration Levels. The other is the subjective workload assessment technique [14] which describes the operator workload in three dimensions: Time Load, Mental Effort Load and Psychological Stress Load. Some of the dimensions have been considered in both techniques [15]: the Time Load and Temporal Demand dimensions; the Mental Effort Load and the Mental Demand and Effort dimensions; and the Psychological Stress Load and Frustration dimensions. Both techniques are largely used in the field of aeronautics, where Collet, Averty, and Dittmar [16] found a positive correlation between the number of aircrafts to control and the NASA-TLX score, amidst traffic controllers, indicating a high sensitivity of NASA-TLX to small workload changes.

Subjective Functional State Measures. According to Thayer [17], task performance is sensitive to the individual functional state, highlighting that arousal state underlies the behaviour. Therefore, the model of multidimensional activation [18, 19] is composed of two dimensions, the energetic arousal and the tense arousal. The tense arousal is considered to be determined by danger, and to be largely cognitively mediated. Whereas the energetic arousal varies naturally according to circadian rhythm, and as a function of factors such as: time of the day, exercise, nutrition, and mental workload [17, 20]. Each dimension level is relative to the others, because these are assumed to form a curvilinear relationship. That is, these are positively correlated at low levels, and negatively correlated at high levels [21]. Therefore a phenomenon considered dangerous has a greater psychological impact (leading to a greater tense arousal) when the energetic level is low; and a lesser impact when the energetic arousal is high. Consequently, considering these dimensions (tense arousal and energetic arousal) during a critical situation seems essential. The Short Form Activation-Deactivation Adjective Check List (AD-ACL) can be used to assess energetic and tense arousal [19]. This is composed of four subscales: energy, tiredness, tension, and calmness. The ratio between the energy and tiredness scores is employed to assess energetic arousal and the ratio between the tension and calmness scores, to assess tense arousal.

Subjective Emotion Measures Scherer [22] proposed a Component Model of Emotion (CME), by considering emotion like an episode of interrelated and synchronized state changes, resulting from the situation evaluation. The model components are: cognitive appraisal, physiological reactions, behaviour tendencies, motor expression, and subjective feeling (emotional experience). Two of this model's components: cognitive appraisal and emotional experience can be estimated by subjective measures.

Cognitive Appraisal consists of a situation evaluation under direct emotional responses (positive or negative). Demir et al. [23] consider the following appraisal components: consistency of motives, intrinsic pleasure, expectation confirmation, standard conformance, agency, coping potential, and certainty. Scherer [22] proposed the Geneva Appraisal Questionnaire (GAQ) to assess the result of an individual's appraisal process during a specific emotional episode. This tool allows estimating five dimensions: intrinsic pleasantness, novelty, goal/need conduciveness, coping potential and norm/self-compatibility. The emotions taken into account by GAQ are: anxiety, irritation, contentment, joy, sadness, disgust, fear, anger, and surprise.

Emotional Experience is characterised by emotion type and intensity. Two tools can be used to measure it: the Geneva Emotion Wheel [24] and EMOTAIX [25]. GEW is a verbal self-reporting instrument on which an individual is asked to indicate the emotional intensity felt in a particular situation, amid 20 emotion categories, such as: interest, irritation, contentment, joy, sadness, disgust, fear, anger, and surprise. Five degrees of intensity are available for each emotion category, and these are represented by circles of increasing sizes, as a function of the emotional intensity. EMOTAIX, in turn, allows analysing the emotional lexicon, used by an individual, when reporting the feelings experienced in a situation. This is a computer based application that works with corpus analysis—Tropes software (version 7). This tool enables to account for the emotional lexicon (2.014 references) according to hedonic dimensions (positive or negative valence) and according to 28 basic thematic categories, grouped in super and supra categories.

6.2.2 Performance Measures

Performance is the result of an established behaviour, more or less adapted to a situation, and corresponds to response accuracy and response latency. Chi and Lin [26] demonstrated a trade-off between these performance criteria. The time needed to complete a task increases when accuracy requirements increase, whereas a decrease in accuracy occurs when task speed requirements increase. In the same line, Fournier et al. [27] proposed a method to arrive at a performance index, taking into account both response accuracy and response latency. They evaluated subjects' behavioural responses in a multi-task context by calculating a composite standardized Z-score for each subject. For each task, the ratio between the reaction time and the proportion of correct responses was weighted by one-quarter and, the correct ratios were summed up. Results revealed that the global performance decreased as the task demand increased and, that the performance improved with training, especially in high task demand conditions.

6.2.3 Psycho-Physiological Measures

The managing of a critical situation causes changes to the autonomic system which consists of two systems: one controlled directly by the nervous system and another controlled by the hormonal system (adreno-medullary). These two systems have different functional roles [28]. Whereas the former executes precise, rapid, and often highly differentiated adjustments, the latter independently modifies important metabolic functions. The two systems may mutually support each other, when massive and generalized system activation occurs, as it is the case during a critical situation. Measures commonly employed to evaluate autonomic system demand are respiratory, cardiovascular and electro-dermal measures. One major advantage of using physiological measures is the continuous availability of bodily data, allowing the reactions to be measured at a high rate and with a high degree of sensitivity, even in situations when open behaviour is relatively rare [29]. Physiological measures are also very sensitive to physical effort and, will reflect specific mental load or emotional variations for activities involving little or no physical effort [30]. The measurement of cardiac activity is a physiological technique employed in the assessment of mental workload and emotional aspects (i.e. anxiety). It has been demonstrated that the heart rate variability (HRV) [31] shows a systematic and reliable relationship with task demand [32, 33]. Similarly, electro-dermal measures are being used to estimate emotional feeling such as anxiety, contentment, joy and fear [34], and it has been found that these are very sensitive even to low emotional changes.

The tools presented in this section were conceived to estimate an individual's reactions in several levels. However, these levels are not parallel, but nested within each other. A study by [35] highlighted the relationship between the indicators to be adopted in this study: (1) subjective measures (self-rated effort; energetic arousal), (2) performance measures (correct responses), and (3) physiological measures (heart rate variability). The model presented on this chapter was based on this study, which suggests that the measures displayed differential sensitivity to the three contextual factors investigated, and that some measures can be determinants of others. Specifically, the heart rate variability increased with high energetic arousal, whereas the participants' self-rated the mental effort as sensitive to both: task difficulty and time pressure. Performance was determined both by energetic arousal and by the interaction between task difficulty and time pressure. Thus, the only satisfactory solution to apprehend the human behaviour in the critical situations is to diversify metrics hoping to achieve a more comprehensive view of the potential repercussions of the contextual factors on an individual, at all levels.

6.3 A Simplified Abstract Model of the Human Behaviour

During this research an abstract model was conceived to represent a set of variables and their relationship, aiming to support the understanding of the human behaviour during critical situations. The model includes six major components, each composed of elements of interest to investigate. These are represented in Fig. 6.1: (i) an individual functional state, based on his or her state of alert state and anxiety level; (ii) an individual characteristics (profile, personality and attention); (iii) situation characteristics (stressing elements, tool's usability and task to be performed—test scenario); (iv) the situation awareness through workload and emotional state, both as perceived and expressed by the individual; (iv) decision making (situation cognitive evaluation) and (v) the achieved human performance is expressed in terms of: task duration; number of errors and final task state.

The relationship between components is defined by the influence of one component over another. For instance, the situation awareness is influenced by: the individual and situation characteristics and the individual's functional state. Within situation awareness, the workload and the emotional state influence each other. The performance is influenced by the decision making, and both depend on the situation awareness.

It must be pointed out that other variables, also accounted for in the simplified model are equally important for the understanding of the human behaviour, in spite of not being discussed on this chapter. These are: the individual's characteristics such as profile (age, gender, and schooling, knowledge on the system, the task and the work context); personality trait, attention and concentration levels and, situation characteristics such as usability of tools and systems used to perform the task, stressing events and decision making style, based on the cognitive evaluation of the

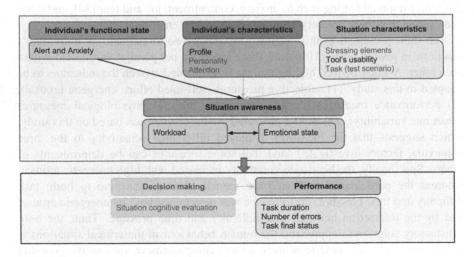

Fig. 6.1 The abstract simplified model of the human behaviour

situation. It must be highlighted that the scenario devised for the experiment concerns a specific task and a set of stressing factors to be presented to the individual taking part in the experiment along its course. Therefore, in spite of not being discussed in this chapter, the case study chosen scenario accounts for these variables and their relation with the user behaviour.

Cognitive resources are essential to an individual decision making process, and in this context these determine task performance. Their availability is determined by the individual functional state, represented in the model as alertness and anxiety levels. Cognitive resources consumed before task execution are spent with the individual perception of the situation in terms of the workload; and with the individual's emotional state. Situation awareness is influenced by the individual's and situation characteristics. The remaining cognitive resources can be obtained by subtracting the consumed resources from the available resource. The result is left for decision making which in turn influences an individual's performance.

In the model, human performance is characterized by the variables: duration of the task, task completion state and error rate, all measured during task performance. These variables are influenced by the user perception about the experiment itself, characterized by the inter-relationship between workload and emotional state. Whereas situational awareness is the result of the following individual characteristics: profile, functional state of alertness and anxiety, as well as situation characteristics coupled with the usability of the tools employed to perform the task.

The following section introduces the experimental protocol conceived to support the understanding of the human behaviour based on the above model. It guides the process of experiment planning and data gathering during the observation.

6.3.1 The Experimental Protocol—PEOI

In the human interface domain, the usability evaluation practice is based on the observation of individuals during task performance. This practice allows identifying the elements of the interface which prevent goals achievement as well as the potential solutions to the perceived problems. Since it also consists in observing human behaviour, the data gathered might be biased by the awareness of the observed individual. Therefore it is paramount to follow a protocol in all of the experiment's phases: planning, conducting, analyzing and reporting the results, in order to ensure that consistent data will be available. Such detailed description of procedures and activities supported by specific documents compose an experimental protocol [36].

This section briefly describes the experimental protocol adapted to support the observation and gathering of human-behaviour data from the interaction, during critical situations. The Experimental Protocol to Observe the Interaction (PEOI) reflects the model conceived to support the understanding of the human behaviour, presented above. Its application is supported by tools employed in the observation of the human behaviour in a controlled environment during the simulation of critical situations.

PEOI's structure is based on the principles adopted during the observation of the human behaviour in the areas of product usability testing and work ergonomics in cognitive psychology. It originates from the practices adopted in the Human Interface Laboratory—LIHM, at the Federal University of Campina Grande (UFCG), in Brazil, and resulted from the compilation of twenty seven years of product usability testing experiences at LIHM, refined with contributions from the literature review on usability evaluation practices, such as those described in the work of Nielsen [37]; Mayhew [36]; Redish [38]. It is structured in phases, goals and artefacts. Its original version describes how to plan, execute and report on the observation of the interaction between users and products [39].

In order to adapt it for the observation of the user behaviour, the original version was reviewed to: (1) establish the relationship between phases, processes and activities; (2) define the roles of the actors involved in the experiment; (3) propose tools for data gathering and analysis, which can be adequate to handle the variables of interest for the human behaviour model.

The protocol's process flow is organized in six phases, each one with a clear objective, and consisting of <u>processes</u> detailed into <u>activities</u>:

Phase 1 Experiment Planning, which is crucial for the correct observation

Phase 2 Participant's training. This phase is optional, depending on the observation goals

Phase 3 Experiment Elaboration and Validation. During this phase, the team must organize all the materials to be used during the experiment

Phase 4 Driving the test and Data gathering

Phase 5 Data preparation and Analysis. This phase consists of organizing the data gathered in the previous phase: questionnaires, interviews, audio and video recordings, as well as physiological data recordings. This is followed by the analysis of the situation characteristics and participants' functional state, decision making and performance

Phase 6 Presenting the results as a diagnostic containing the findings on participants' behaviour

An ensemble of roles and responsibilities are defined for all experiment participants. A role represents a set of responsibilities assigned to a participant. The roles were grouped into classes and represented in the protocol as a workflow.

To support the protocol application during all phases of an experiment, a set of artefacts and respective templates is available to the experiment's team. The artefacts are grouped according to the phases: planning, application, data analysis and reporting on results. Some artefacts were conceived to ensure the ethics of the experiment's procedures, including terms of agreement with the experiment conditions whereas others are terms of confidentiality over the tools and task procedures employed during the experiment and which belong to the stakeholders.

6.3.2 PEOI: Data Gathering Methods

Data gathering in the protocol consists of a combination of methods: interaction observation; interviews; questionnaires; physiological measurements; document analysis and video and discourse analysis. The choice of methods to be combined is a function of the observation goals and the phase of the protocol. The tools proposed for data gathering are illustrated in (Fig. 6.2). These are grouped into four categories: individual's functional state; individual's characteristics; situation characteristics; situation awareness; decision making and performance.

The tools' categories reflect the components in the Human Behaviour Model. Given the focus of this chapter, it follows a brief description of only the subset of tools applied in the case study described in Sect. 4.

Individual's characteristics: individual's profile data is obtained through the application of the questionnaire: User Objective and Cognitive Profile (POCUS) [40]. It gathers information on: personal, physical, professional, contextual, psychological and clinical aspects of an individual.

Situation characteristics: product usability level is measured by applying the questionnaire Webquest [41]. The usability level is based on an individual's subjective satisfaction with a product (tool) employed to do a task during the observation experiment. This questionnaire investigates the following aspects: navigation and product use, documentation (and help mechanism), product feedback in response to its user actions and product-user interaction. The result is a satisfaction index level in the scale: extremely satisfied (0.67−1), very satisfied (0.33−0.66), marginally satisfied (0.01−0.32), neutral (0), marginally dissatisfied (−0.01 to −0.32), very dissatisfied (−0.33 to −0.66) and extremely dissatisfied (−067 to −1).

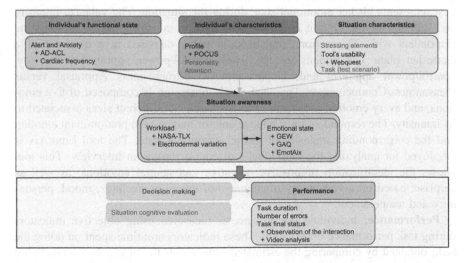

Fig. 6.2 Data gathering tools

Functional state: to investigate the anxiety and the alert levels in an individual, two tools are employed: the questionnaire Activation-Deactivation Adjective Check List (AD-ACL), which allows evaluating the transitional state of alertness. To this transitional state it is assigned one of the levels: general activation, sleepiness, high activation and general deactivation. With this knowledge it is possible to determine: the actual state of alertness (arousal energetic); and the state of tension (tense arousal) which combined determine the levels of anxiety and alertness. The second tool, measures the heart rate variance, to evaluate the anxiety level. In this case, the anxiety level is extracted from the interpretation of the spectrum, in the fourth band of frequency: very low (0.01−0.04 Hz); low (0.04−0.15 Hz), high (0.15−0.40 Hz) and Ultra High (10^{-5} a10^{-2} Hz). The relation: low frequency/high frequency, allows evaluating the sympathovagal balance [42]. When the balance is found, the anxiety is discarded; this happen when the results present a variation of ± 2.33. According to Sztajzel [43] this variation must be approximately 4.61, for healthy individuals.

Situation awareness: Two dimensions are assigned to an individual perception of a situation: the workload perception and the emotional state. To measure the workload, PEOI proposes to apply two tools: the questionnaire NASA-TLX, and the electrodermal variance (physiological measurement) during task performance. The NASA-TLX questionnaire identifies the individual perception on his/her workload. It covers different aspects of the workload, such as: behaviour (effort and performance); task demand (mental, physical and temporal); and subjective (frustration). If the sum of all aspects is higher or equal to 60 points, the individual is characterized as being within the acceptable workload level. In order to evaluate the emotional state, PEOI proposes the tools: GAQ and GEW questionnaires and a discourse analysis tool—EmotAix. GAQ investigates the relevance, implications, potential for coping and the compatibility of a situation with norms and standards; considering it all essential to know the user cognitive evaluation ability. This questionnaire presents the respondent with a list of emotions to select the one that best represents his/her emotional state during the situation. The valance, intensity and duration of the chosen emotion must also be informed. GEW presents the respondent with 20 emotions organized in pairs, disposed at a quadrant of a Cartesian plan. The dimensions are: High control/power appraisal versus low control/power appraisal and, Unpleasantness/Obstructiveness Appraisal versus Pleasantness/Conduciveness Appraisal. Each dimension is composed of five emotions, and every emotion is associated to five circles of different sizes associated to its intensity. The respondent must choose one (or maybe two) predominant emotion and the corresponding intensity of his/her emotional state. The tool EmotAix is employed for analyzing the individual's discourse during an interview. This tool allows the identification of positive, negative and neutral emotions, as well as surprise; based on a vocabulary (dictionary) of emotions, feelings, mood, personality and temperament.

Performance: individual performance is measured using objective indicators during task performance observation. These indicators are: time spent on doing the task; obtained by comparing the estimated time versus the time actually taken; the number of human errors incurred when performing the task; and the task state at completion, that is task completed with or without fault or task unfinished, abandoned or interrupted. These data are gathered through direct observation of the variables of interest. The observation can be enhanced by video and audio recordings.

Besides the described tools for data gathering, the protocol also provides guidance on how to analyze the collected data, as will be shown next.

6.3.3 PEOI: Data Analysis

In the protocol, data analysis is performed in two steps. In the first one must select how to analyse each set of data, according to the tool used for gathering. In the second step a data analysis correlation is performed. The objective of the latter is to correlate data across the categories of the proposed Human Behaviour Model. During this second phase the variables are separated into two groups, internal and external variables, according to their pattern of cognitive resources consumption [35].

6.4 Case Study

This section presents a case study to illustrate the Human Behaviour Model application, and the supporting protocol (PEOI). Its main objective was to allow investigating the relevance of the collected data and analysis. The situation consisted in observing and analyzing the user behaviour when interacting with a product during a critical situation—a crisis management, in a simulated environment. During the experiment the participants were asked to generate an intervention plan after a maritime accident, within a strict time constraint. The plan generation was supported by the decision making aid tool—*Generateur de Plans d'Intervention* (GENEPI) [44]. The experiment consisted in simulating the situation, inspired in a report of a real crisis. The protocol PEOI was employed to observe the user interaction with the system GENEPI. The aim was to understand the relationship between the human error and the participant's behaviour. The data was gathered though observation, video recording, questionnaires, interviews and with the aid of specific tools to gather physiological data.

In this case study only a subset of the model's components was considered, which were chosen to investigate the participant's: profile; functional state; situation awareness; and the task tool usability and, objective indicators linked to the individual's performance. Therefore only a subset of the variables present in the Human Behaviour Model was explored. The variables of interest were: (1) the experiment participant profile, gathered with POCUS; (2) the participant functional state, considering the levels of alertness and anxiety as measured with AD-ACL and the heart rate variance; (3) the situation awareness based on the workload (measured with NASA-TLX and variance electrodermal); (4) the participant emotional state (measured with GEW, GAQ and EmotAix); (5) the subjective participant

Table 6.1 Knowledge and expertise on: product use, task and, working context

Participant		P1	P2	P3	P4	P5	P6	P7
Knowledge and expertise	Product	Low	Low	Low	Low	High	High	Medium
	Task	Low	Low	Low	Low	High	High	Medium
	Context	Medium	Low	Medium	Low	High	Low	Low
Test scenario			Open	Guided	Guided	Guided	Open	Open	Guided

satisfaction with the working tool assessed through product usability and situation characteristics, assessed with Webquest; and (6) the performance variables during the experiment, obtained through direct observation and video analysis. Details of the experiment will be given in the following subsections.

6.4.1 The Experiment

Place, Time and Sessions: The experiment was performed in 2011, at the Research Centre for Knowledge Psychology, Language and Emotion,[1] throughout seven sessions including a pilot one.

Test participants basic profile: the group of seven participants consisted of three men: P1, P5, P6 and four women: P2, P3, P4 and P7. These spanned a large age group: P2, P3 and P6 aged between 18 and 24; P1and P7 aged between 25 and 35, and P4 and P5 aged above 35. The participants' academic background was: P4 and P5 Ph.Ds; P2, P3 and P6 undergraduate students; P1 a Master student and; P7 a doctorate student. The participants' academic domain was risk analysis, computer studies and human interface studies. Their level of expertise on the domains: product (GENEPI) use; task to be performed during the experiment (generating contingency plans) and; on the working context (risk management), is displayed on Table 6.1.

The task was organized into two scenarios, which differed on the level of details given to the participants, and on the time allocated for task completion. The guided scenario was given to participants with lower levels of expertise, to be completed within 30 min (estimated time). The open scenario, with less guidance to the participants, had an estimated completion time of 40 min. The difference in completion time aimed to balance the cognitive demands on both groups of participants.

Product: GENEPI is a decision making support system conceived to assist crisis management, which supports the elaboration of contingency plans for maritime accidents. It is one module of a tool developed for crisis management related to marine pollution. Its objective is to facilitate and accelerate the establishment of exclusion zones, and to mobilize the appropriate means to handle these critical situations.

---

[1] Le Centre de Recherche en Psychologie de la Connaissance, du Langage et de l'Émotion (Centre PsyCLÉ) .

Participants training: The participants had an introductory presentation on the use of GENEPI's functionalities and user interface given by the development team, which lasted 1 h and 30 min.

Task and interaction Context: The experiment emulated a crisis scenario which was inspired on maritime accident reports from real situations. During this scenario the participant was asked to use the tool GENEPI to generate a contingency plan for the described maritime accident, also known as an intervention plan. The work context in which the participants were immersed was prepared in order to emulate that at the *Centre Opérationnel de Surveillance du Littoral* (in the French Mediterranean region), which is the first organism contacted after an accident happens. The work scenario begun with the participant receiving a simulated phone call from the *Préfecture Maritime de la Méditerranée*, informing about the accident. The severity of the simulated accident required that the intervention plan should have been generated within time restrictions imposed by another simulated phone call from the *Préfecture*. The participant was advised in the task description text to get additional information on the accident situation by contacting specialized services through phone calls using a given phone directory. The phone contacts were simulated with the aid of supporting participants who performed predefined roles.

Experiment team: consisted of a group of six people with multidisciplinary skills who performed multiples roles, amidst interacting with the participant during the experiment. Two of the team members were product usability experts; one was an expert in cognitive psychology and, two undergraduate students in computer science.

6.4.2 Data Gathering and Analysis Results

The data was gathered **in accordance with the** experimental protocol PEOI. It was collected during the experiment was classed and analysed for each model variable. In order to evaluate the behaviour model, four hypotheses were investigated in this study:

- H1: Individuals who were more dissatisfied with product usability (Webquest) tend to express more negative emotions (EmotAix);
- H2: Individuals tend to display higher alert levels (AD-ACL) when they perceive a situation of work overload (Electrodermal peaks);
- H3: Individuals tend to display higher anxiety when they perceive a situation of work overload (Electrodermal peaks);
- H4: Individuals tend to have a higher error rate when they are under negative emotions (EmotAix).

The analysis was performed in two steps. The first step consisted in visually investigating a correlation between the model variables, which was performed by the domain specialists in usability and cognitive psychology. This was followed by a exploratory data analysis performed with the same purpose.

The exploratory data analysis, initially investigated the normality of the data distribution using the Shapiro-Wilk test [45, 44]. A significance level of 5 % (0.05) was adopted, with a level of confidence of 95 % that variables with the p-value higher than 0.05 follow a normal distribution. This was followed by a correlation investigation between sets of variables, using the Spearman's rank correlation coefficient [46]. Dispersion graphs were then built in order to propose an appropriate regression model for the relationship between predicting variable and response variables (linear or non-linear), for each hypothesis. As a result it was determined the influence of the predictor variable over the response one. It follows the analysis results.

Situation characteristics: GENEPI usability was indirectly measured through a user satisfaction Index. Table 6.2 presents this index calculated for all the experiment participants, according to their individual subjective satisfaction level. The opinion of the majority of participants (5/7) resulted in a negative index, which represents a moderate dissatisfaction (P7, P2, P3, P4 and P1). The other 2 participants (P5 and P6) displayed a high satisfaction level. This contrast may have resulted from the difference in participants' familiarity levels with GENEPI; with the highly satisfied users being the most familiar ones.

Individual's characteristics: these were self-assessed by the participants who answered a questionnaire. According to their answers, three participants: P5, P6, P7 considered their abilities for learning and knowledge application as high, three participants: P1, P3 and P4 self rated as average, and one participant: P2 did not answer this question. Regarding computer literacy, except for one participant: P4 (declared as average), all the others declared as high. Regarding a set of specific psychological characteristics, it follows the distribution. Ability to analyze situations and to solve problems: P1, P2, P3, and P7 self assessed as average, whereas participants P4, P5, P6 self-assessed as high. Regarding the sense of direction, P2, P3, P4 and P7 declared as average, whereas P1, P5 and P6 self-assessed as high. The distribution of self-assessment regarding the level of abstraction was: P4, P5, P6 and P7 declared as high, whereas P1, P2 and P3 declared as average. Concerning organization and planning: P1, P3, P4 and P7 declared as high, whereas P2, P5 and P6 declared as average. Lastly, regarding the ability to have a wide panoramic view of a particular situation, participant P5 did not answer this question; participants P1, P2, P3and P4 self assigned an average level, whereas participants P6 and P7 declared a high level.

Table 6.2 User satisfaction levels with GENEPI

Participant		P1	P2	P3	P4	P5	P6	P7
Satisfaction level	Index	−0.068	−0.175	−0.171	−0.126	0.488	0.412	−0.240
	Moderate dissatisfaction	X	X	X	X			X
	High satisfaction					X	X	

Functional state (alertness and anxiety levels): the questionnaire AD-ACL was employed to determine this state, with its four dimensions: activation general, inactivity/sleepiness, high activation and inactivity general. For each dimension, the maximum score is 20 points. All participants scored high, showing that they were alert and aroused, according to the measured *activation general*. Consistently, the *inactivity and sleepiness* levels were low for all users. Given the *high activation*, the users were considered not tense during the experiment. According to the *inactivity general*, the participants were in a medium level of tranquillity. This data was converted into levels of alert and anxiety for each participant, with the maximum score being 4 points, as shown in Fig. 6.3. When the obtained index (shown in the abscissa of Fig. 6.3), is above 2, then anxiety and alertness are present. Two participants were anxious (P4 and P6) and five participants (P5, P1, P3, P6 and P2) were alert.

From the analysis of the heart rate variation, with reference values in the range of 2.28 and 6.94, participants with an index greater than 6.94 were classed as having a significant anxiety level. Therefore, as shown in Fig. 6.4, anxiety was detected using physiological measures in four participants: (P1, P3, P4 and P7), with participant P4 displaying the highest value.

Situation awareness (workload and emotional state): In the experiment, the perceived workload was measured by the questionnaire NASA-TLX. The results show that three participants scored above the line which divides the non-excessive and excessive workload perceptions. Thus three participants (P2, P4, P6) considered it excessive; whereas four participants did not consider it excessive (P1, P3, P5 and P7—see Fig. 6.5).

The NASA-TLX questionnaire also allows identifying the dimensions that most influenced the workload perception. The results showed, in decreasing order of relevance, the following dimensions: mental demand, frustration, cognitive effort, temporal demand, performance and physical demand. Therefore, the task represented a high cognitive workload, combined with a considerably high level of effort, leading to participant frustration.

In the experiment, the workload was also measured by electrodermal activity (EDA). The strategy adopted was to identify the number of peaks (independent of amplitude) and associate those to the situation context (events, stress factors,

Fig. 6.3 Anxiety and alert levels

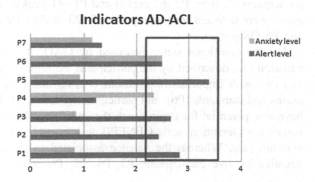

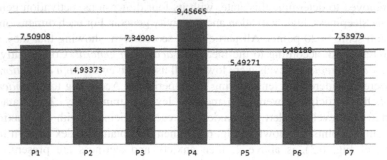

Fig. 6.4 Participants' Sympathovagal balance (anxiety)

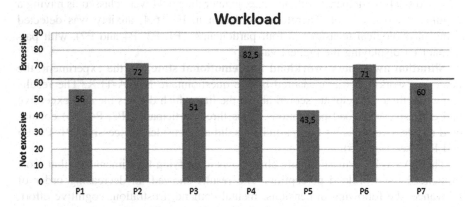

Fig. 6.5 Workload perception

section of task under execution, etc.). The aim was to interpret the electrodermal variation in the context of its occurrence. The number of peaks suggests a workload level to be investigated. The greater the peak occurrence the higher the workload to which the participant is subjected. The higher number of peaks was detected for participants P5 (65), P2 (62 peaks) and P1 (41 peaks), whereas lower numbers of peaks were detected for participants P7 (23 peaks), P3 (25 peaks), P4 (29 peaks) and P6 (38 peaks).

For the emotional state, measured by GAQ, it is important to note that the situation was described by the participants as: highly relevant (P1, P2, P3, P4, P5 and P6); with important implications (P2, P3 and P7); low in compatibility with norms and standards (P6); and participants (P1, P3, P4, P5, P6 and P7) declared to have low potential for coping with the situation. Therefore, one can state that the situation (interaction with GENEPI) was considered by the participants as an extreme case. Whereas the valence associated to the situation was perceived as negative by five participants (P2, P4, P5, P6 and P7), reaching the maximum intensity for two of those (P4 and P5). The emotions chosen by participants to describe their emotional state were: anxiety, irritation and sadness. The last finding with GAQ was the participants' attempt to minimize or mask their feelings during the experience. Regardless of positive or negative valence, all participants tried to reduce the intensity and duration of their emotional episodes and tried to control or mask their feelings.

With GEW, it was possible identifying the valence and the intensity of emotions. The negative emotion was displayed by five participants (P2, P3, P5, P6 and P7), two of which with the maximum intensity (P6 and P7). Participant (P4), amidst the two who displayed positive emotions, reached the maximum intensity. It must be pointed out that the two participants with positive emotions were part of GENEPI development team, therefore they knew it well and were positively biased in their evaluation.

EmotAix was employed in the analysis of discourse recorded during the interviews after the experiment. The results showed that for all participants prevailed negative emotions (Fig. 6.6).

Performance: In the experiment, performance was measured in terms of a set of dependant variables, one of which was the relationship between the estimated time to perform the task and the actual time taken to perform it. The estimated time for guided and open execution scenarios were respectively 30 and 40 min. This relationship is represented in Table 6.3. Given that only one participant (P1) did not exceed the estimated time, it was concluded that the estimation was not adequate.

Another indicator of performance in the model is the final state achieved through performing the task, that is: *task finished with or without faults*; *unfinished due to interruption* or *abandonment*. Six users did not finish the task (P1, P2, P3, P4, P5 e P7). Half of those interrupted (P2, P5, P7), and half abandoned (P1, P3, P4). One of the reasons for abandoning the task was reaching the estimated time limit; which lead the team conducting the experiment to interrupt the task. Another reason was

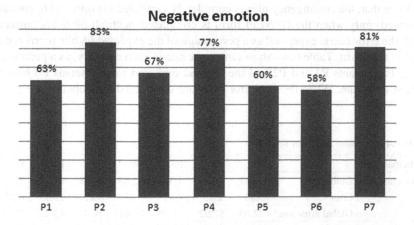

Fig. 6.6 Participants' negative emotion detected by EmotAix

Table 6.3 Estimated versus execution time

Participant	P1	P2	P3	P4	P5	P6	P7
Estimated time	40′	30′	30′	30′	40′	40′	30′
Execution time	29′32″	41′24″	32′58″	33′01″	43′34″	51′08″	34′18″
Over time	–	11′24″	2′58″	3′01″	3′34″	11′08″	4′18″

due to problems with the version of GENEPI, making it difficult or impossible to finish the task; and the unavailability of the necessary data to complete the task. Only participant P6 finished the task within the time limit, but this was achieved with faults.

In this study the human errors were classed in two groups. The first group was related to the interactive process, and consisted in not filling one or more fields in the GENEPI form, or filling incorrectly. The second group of errors consisted in failing to obtain the necessary information about the incident, in order to complete all GENEPI's forms required to generate the intervention plan. This data acquisition consisted in phoning as series of organisms to obtain the required information. The authors consider that both kinds of errors were directly related to time pressure and distractions, some of which imposed by the experiment team in other to generate observable stress levels.

The failure to obtain the required information to fill the forms not only influenced the time taken to complete that task but also the task outcome and the participant's emotional state. Given this correlation between the model variables the analyzed errors are those in the group 1, that is, errors in the interactive process. Thus, the global error rate is displayed in Table 6.4, with a highlight for the highest errors rates.

GENEPI consists of 68 fields organized in seven forms or tabs. To complete the task successfully, the user must fill in all fields correctly. Interaction between the user and this tool results in: (i) Omissions—fields left incomplete; (ii) Errors: fields populated with wrong data, and (iii) Success- fields correctly filled.

Given that, the contingency plan is completely generated (as opposed to partially generated) only when the GENEPI form is completed, with all 68 fields correctly filled, the error rate is expressed as a percentage of the expected 68 hits (correct data filled in the form). Table 6.4—shows the error rate in each category, as a percentage value. Participants P4 and P2 had the highest omission rates whereas P6 had the highest error rate. When the two error rates are considered, one obtains the global

Table 6.4 Omission and error rates (%)

Participant		P1	P2	P3	P4	P5	P6	P7
Error classes	Omission	16.17	44.1	7.35	72.03	20.58	0	17.64
	Error	5.88	8.82	4.41	5.88	8.82	13.23	8.82
	Global error rate	22.05	52.92	11.76	77.91	29.4	13.23	26.46
Classification		5°	2°	7°	1°	3°	6°	4°

error rate, for which P2, P4 and P5 showed the worst performance. The table also rates all participants in terms of performance, according to their error rates, in decreasing order.

6.4.3 Results Discussion

The previous section presented the results obtained with the protocol application when collecting the information required by the Model of the Human Behaviour in order to investigate the correlation between the model variables. The correlation could help explain the reasons for the human error occurrence, during the interaction with systems developed to support tasks under critical situations. Employing a variety of tools when gathering the same kind of data, aimed to investigate the best suited ones for the purposes of this research—that is to investigate the likely causes of human error during interaction.

A summary of the data collected in the experiment is shown in Fig. 6.7. This Figure allows exploring the relationship between the variables investigated in the Model of the Human Behaviour, and adopts the same variables representation as the model. This global view of the data allows performing analyzes under different points of view. In the figure, the data collected is organized into categories defined in the model: individual's functional state; individual's characteristics; situation characteristics; situation awareness and performance; and grouped according to the variables of interest: alert; anxiety; participant profile; usability satisfaction; test scenario; workload perception; emotional state; task duration; global error rate and task outcome.

The variables and their respective value domains were represented in iconic form, as explained in the legend. The sequence adopted when representing the participants starts from left to right with P1–P7. This representation allows both individual and group analysis of the investigated variables.

Figure 6.7 also makes reference to the instruments used to collect each variable's data. In some cases, more than one tool was used to collect the same variable. For instance, workload perception was captured using NASA—TLX and electrodermal variation. During this study it was intended to confront results obtained through participant opinion with stimuli involuntary responses. Differences in results can be explained through individual characteristics, such as personality traits, which influence the situation perception as positive or negative whereas physiological data is independent, since it is involuntary. Therefore, the authors recommend gathering independent data at the beginning of the experiment to use as reference values (base values) against which variable variations along the experiment can be identified.

It follows an example of analysis with the purpose of helping to understand how to interpret the information content of Fig. 6.7. The analysis is simple and does not exhaust all possibilities of analysis of the data there represented. The analysis consists in exploring with the aid of the model, the potential causes behind the three cases of lowest performance (global error rate) individually presented by the participants: P2, P4 and P5.

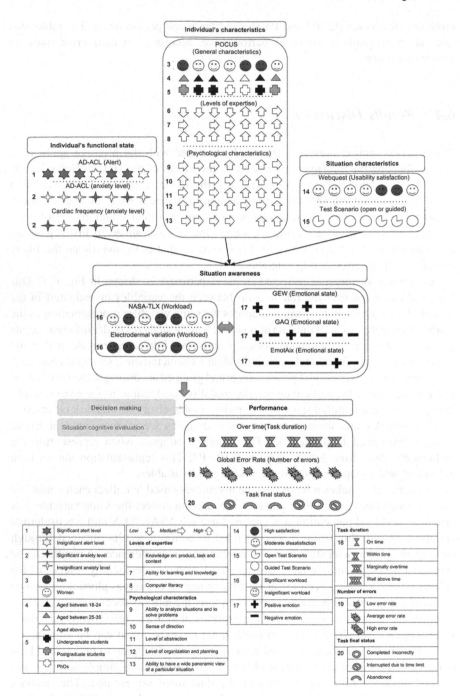

Fig. 6.7 Experiments' results mapped into the human behaviour model

P2—presented the highest global error rate in the group. The combination of low anxiety levels: declared though AD-ACL, and measured (cardiac frequency); with high declared alert level (AD-ACL), low level knowledge on the task, tool and context and moderate dissatisfaction with GENEPI usability, contributed to the high workload perception declared though (NASA-TLX), The electrodermal variation was compatible with felt and expressed negative emotions (GEW, GAQ e Emo-tAix). This combination of variables could explain the high error rate and the decision of interrupting the task (guided) soon after the estimated time of 30 min.

P4—presented the second highest global error rate in the group. The high values obtained with the declared anxiety (AD-ACL) and that measured physiologically (cardiac frequency) combined with the low alert level (AD-ACL), coupled with the low knowledge on the task and its context, and the moderate dissatisfaction with GENEPI's usability, contributed to the high workload perception (NASA-TLX) leading into negative emotions (GAQ e EmotAix). This combination of variables might explain the high error rate and the decision to abandon the task (guided) immediately after reaching the estimated time of 30 min.

P5—with the third highest global error rate, this participant presented low anxiety levels, both declared (AD-ACL) and measured (cardiac frequency), combined with a declared high level of alertness (AD-ACL). Contrasting with participants P2 and P4, this participant's knowledge on the task, the tool and the context was high, therefore he declared himself satisfied with the tool's usability. His situation awareness, both declared (NASA-TLX) and measured (electrodermal variation) was a negative one—high workload, resulting in expressed negative emotions (GEW, GAQ e EmotAix). This combination of variables resulted in a high error rate and the task outcome—interrupting the task (open) incomplete, after having exceeded the estimated time of 40 min. In spite of his high knowledge, high satisfaction with the tool and low level of anxiety, time pressures coupled with stressing factors introduced during the test resulted in a similar negative perception of the situation as the previously described participants.

Therefore, it could be inferred of the above analysis results, involving the three participants (P2, P4 and P5) that time pressures combined with stressing factors are very influential variables over the task outcome and the global error rate.

It must be explained that the stressing factors employed during the various sessions were different in each case and, these were chosen according to the previous knowledge of the participants' profiles, aiming to cause a significant impact on the user attention and stress levels to be detected by the measuring instruments. An example of a stressing factor is constant interruptions by phone calls during the task. On the other hand, different stressing factors may also have caused different stress levels among the participants, interfering with the task outcome.

Although the analysis results can contribute to the understanding of aspects of the individual behaviour, it does not support the extraction of a standard human behaviour or supports inferences. This study demonstrated the complexity of trying to correlate the model variables, which remain a challenge to be further investigated. Based on calculations of the statistical power of the regression test and, aiming to define a sample size capable of reaching a high statistical power of 0,8 and an anticipated effect of (f2 = 0.35); it was found that experiments must be conducted with a minimum of 39 participants (system users).

6.5 Final Considerations and Future Work

This work presented a methodological framework and supporting tools conceived to broaden the information available in the literature on the human behaviour during critical situations and its relation with the occurrence of the human error. Its major contribution is the Simplified Abstract Model of the Human Behaviour, supported by an experimental protocol to support the data gathering in a systematic and organized manner to support the observation of the human activity during critical situations. The simplified abstract Model of the Human Behaviour, accounts for six interrelated components: individual's characteristics, situation characteristics, functional state, situation awareness and performance.

The experiment reported on this chapter enabled to evaluate the quality and utility of the behavioural model in identifying behavioural factors which contribute to the human error during critical situations. It has also shown the viability of behavioural data gathering and analysis using the supporting protocol.

Since it was not possible to infer, a standard human behaviour which precedes the error, the authors recommend further studies, involving a larger sample of participants and broadening the investigations to include all the model's variables and their correlations. This approach should support inferences on which variables in the model are more influential on the user performance.

From this result it is also proposed to weight the influence of each model component on the user behaviour and to integrate this knowledge into the Method for Conceiving Ergonomic Interfaces (MCIE) [47] aiming to reduce the human error incidence due to human interface flaws. Finally, there are plans to build a computer system to analyze the results regarding the knowledge of the human behaviour during critical situations and issue recommendations to improve the interactive process and reduce the human error during critical situations.

References

1. Hollnagel, E. (1993). *Human reliability analysis: Context and control.* New Jersey: Academic Press.
2. Rasmussen, J. (1983). Skills, Rules, and knowledge; signals, signs, and symbols, and other distinctions in human performance models. *IEEE: Transactions on Systems, Man, and Cybernetics, 13*(3), 257–266.
3. Reason, J. (1990). *Human error.* Cambridge: Cambridge University Press.
4. Reason, J. (1997). *Managing the risks of organizational accidents.* London, UK: Ashgate Publishing.
5. Norman, D., & Draper, S. (1986). *User cantered system design in new perspectives on human-computer interaction.* Hillsdale: Lawrence Erlbaum Associates.

6. Norman, E. B., & Baucom, D. (2002). *Enhanced cognitive-behavioural therapy for couples: A contextual approach*. NBR: American Psychological Association. 6023.
7. Endsley, M. R. (1995). Toward a theory of situation awareness in dynamic systems. *Human Factors, 37*(1), 85–104.
8. Amalberti, R. (1996). *La conduite de systèmes à risques*. Paris: Press Universitaires de France.
9. Rasmussen, J., Pejtersen, A., & Goodstein, L. (1994). *Cognitive System Engineering*. New York: Wiley.
10. Rasmussen, J., Pedersen, O. M., Mancini, G., Carnino, A., Griffon, M. & Gagnolet, P. (1981). Classification system for reporting events involving human malfunctions. In: RISO-M-2240. RISO National Laboratory. Denmark. March.
11. van Eekhout, J. M., & Rouse W. B. (1981). Human errors in detection, diagnosis, and compensations for failures in the engine control room of a supertanker. *IEEE Transactions on System, Man, and Cybernetics, 12*, 813–816.
12. Johnson, W. B., & Rouse, W. B. (1982). Analysis and classification of human errors in troubleshooting live aircraft power plants. *IEEE Transactions on Systems, Man, & Cybernetics, 12*(3), 389–393. May–June.
13. Hart, S. G., & Staveland, L. E. (1988). Development of NASA-TLX (Task Load Index): results of empirical and theoretical research. In P. A., Hancock, & N., Meshkati (Eds.), Human mental workload. (pp. 139–183) Amsterdam: Elsevier Science Publisher B.V.
14. Reid, G. B., & Nygren, T. E. (1988). The subjective workload assessment technique: A scaling procedure for measuring mental workload. In P. A. Hancock & N. Meshkati (Eds.), *Human Mental Workload* (pp. 185–218). North-Holland: Elsevier Science Publishers.
15. Miyake, S. (2001). Multivariate workload evaluation combining physiological and subjective measures. *International Journal of Psychophysiology, 40*(3), 233–238.
16. Collet, C., Averty, P., & Dittmar, A. (2009). Autonomic nervous system and subjective ratings of strain in air-traffic control. *Applied Ergonomics, 40*, 23–39.
17. Thayer, R. E. (1987). Problem perception, optimism, and related states as a function time of day (diurnal rhythm) and moderate exercise: Two arousal systems in interaction. *Motivation and Emotion, 11*(1), 19–36.
18. Thayer, R. E. (1985). Activation (arousal): The shift from a single to multidimensional perspective. In J. Strelau, T. Gale, & F. Farley (Eds.), *Biological bases of personality and behavior* (Vol. 1, pp. 115–127). Washington, D.C.: Hemisphere.
19. Thayer, R. E. (1986). The activation-deactivation adjective check list: Current overview and structural analysis. *Psychological Reports, 58*, 607–614.
20. Cariou, M., Galy, E., & Mélan, C. (2008). Differential 24-h variations of alertness and subjective tension in process controllers: Investigation of a relationship with body temperature and heart rate. *Chronobiology International, 25*(4), 597–607.
21. Mélan, C., Galy, E., & Cariou, M. (2007). Mnemonic processing in air traffic controllers: Effects of task parameters and work organization. *The International Journal of Aviation Psychology, 17*(4), 391–409.
22. Scherer, K. R. (2001). Appraisal considered as a process of multi-level sequential checking. In K. R. Scherer, A. Schorr, & T. Johnstone (Eds.), *Appraisal processes in emotion: theory, methods, research* (pp. 92–120). New York and Oxford: Oxford University Press.
23. Demir, E., Desmet P.M.A. & Hekkert, P. (2009). Appraisal Patterns of Emotions in Human-Product Interaction. *International Journal of Design, 3*(2), 41–51.
24. Scherer, K. R. (2005). What are emotions? And how can they be measured? *Social Science Information, 44*, 695–729.
25. Piolat, A., & Bannour, R. (2009). EMOTAIX: Un Scénario de Tropes pour l'identification automatisée du lexique émotionnel et affectif. *L'Année Psychologique, 109*, 657–700.
26. Chi, C. F., & Lin, F. T. (1997). A new method for describing search patterns and quantifying visual load using eye movement data. *International Journal of Industrial Ergonomics, 19*(3), 249–257.
27. Fournier, L. R., Wilson, G. F., & Swain, C. R. (1999). Electrophysiological, behavioural, and subjective indexes of workload when performing multiple tasks: Manipulations of task difficulty and training. *International Journal of Psychophysiology, 31*, 129–145.

28. Folkow, B. (2000). Perspectives on the integrative functions of the sympatho adrenomedullary system. *Autonomic Neuroscience: Basic and Clinical, 83,* 101–115.
29. Paas, F. G. W. C. (1992). Training strategies for attaining transfer of problem solving skill in statistics: A cognitive load approach. *Journal of Educational Psychology, 84,* 429–434.
30. Brünken, R., Plass, J., & Leutner, D. (2003). Direct measurement of cognitive load in multimedia learning. *Educational Psychologist, 38,* 53–61.
31. Backs, R. W. (1995). Going beyond heart rate: Modes of autonomic control in the cardiovascular assessment of mental workload. *The International Journal of Aviation Psychology, 5,* 25–48.
32. Mulder, G., & Mulder, L. J. M. (1981). Information processing and cardio-vascular control. *Psychophysiology, 18,* 392–405.
33. Tattersall, A. J., & Hockey, G. R. (1995). Level of operator control and changes in heart rate variability during simulated flight maintenance. *Human Factors, 37,* 682–698.
34. Kreibig, S. D. (2010). Autonomic nervous system activity in emotion: A review. *Biological Psychology, 84,* 394–421.
35. Galy, E., Cariou, M., & Mélan, C. (2012). What is the relationship between mental workload factors and cognitive load types? *International Journal of Psychophysiology, 83,* 269–275.
36. Mayhew, D. J. (1999). *The usability engineering lifecycle: A practitioner's handbook for user interface design.* San Francisco, CA: Morgan Kaufmann Publishers Inc.
37. Nielsen, J. (1993). *Usability Engineering.* Academic Press, Boston, ISBN 0-12-518405-0 (hardcover), 0-12-518406-9 (softcover).
38. Redish, G (2007, May). Expanding usability testing to evaluate complex systems. *Journal of Usability Studies, 2*(3), 102–111.
39. Aguiar, Y. P. C., Vieira, M. F. Q., Galy, E., Mercantini, J.-M. & Santoni, C. (2011). Refining a user behaviour model based on the observation of emotional states. In: Third International Conference on Advanced Cognitive Technologies and Applications—COGNITIVE, 2011, Rome. Proceedings of the Third International Conference on Advanced Cognitive Technologies and Applications, 2011.
40. Scherer, D. & Vieira, M. F. Q. (2008). Accounting for the Human Error when Building the User Profile. In: Third IASTED International conference human-computer interaction, 2008, innsbruck. proceedings of third iasted international conference human-computer interaction (Vol. 1, pp. 132–137). Zurich: ACTA Press.
41. Vieira, M. F. Q., Queiroz, J. E. R. & Oliveira, R. C. L. (2005). Webquest: a configurable web tool to prospect the user profile and user subjective satisfaction. In: 11th International Conference on Human-Computer Interaction, 2005, Las Vegas—Nevada. HCII 2005. Las Vegas—Nevada: MIRA Digital Publishing.
42. Berntson, G. G., Quigley, K. S., & Lozano, D. (2007). Cardiovascular psychophysiology. In J. T. Cacioppo, L. G. Tassinary & G. G. Berntson (Eds.). Handbook of psychophysiology (3rd edn., pp. 182–210). Cambridge, UK: Cambridge University Press.
43. Sztajzel, J. (2004). Heart rate variability: A non-invasive electrocardiographic method to measure the autonomic nervous system. *Swiss Medical Weekly, 2004*(134), 514–522.
44. Mercantini J.-M., Guerrero C.V.S & Freitas D.D. (2010). Designing a software tool to plan fight action against marine pollutions. In: The 7th International Mediterranean and Latin American Modelling Multiconference, October 13–15 2010, Fes, Morocco.
45. Shapiro, S. S. & Wilk, M. B. (1965). An analysis of variance test for normality (complete samples). *Biometrika, 52,*(3-4), 1965, 591–611.
46. Zar, J. H. (1972). Significance testing of the Spearman rank correlation coefficient. *Journal of the American Statistical Association, 67*(339), 578–580.
47. Vieira, M. F. Q. (2004). Accounting for Human Errors in a Method for the Conception of User Interfaces. In: International Mediterranean Modelling Multi Conference—I3 M'04, 2004, Genoa, Italy. Proceedings of I3 M'04. Bergeggi Italy. *1.* pp. 122–130.

Chapter 7
Hydraulic Dam Safety Assessment with the Timed Observations Theory

Marc Le Goc, Ismail Fakhfakh, Corinne Curt and Lucile Torres

Abstract The safety control process of industrial systems (considered to be dynamic systems) needs to take in account physical processes (e.g. building ageing), informational processes (data collection and processing), decisional processes (data aggregation), and has to consider various constraints (e.g. economic and regulatory). The improvement of informational and decisional processes with the aim of controlling physical processes is based on the development of models and algorithms for measurement, assessment, control, diagnosis and prognostic. In the domain of dam management, assessment of reliability and safety, fault diagnosis, and corrective action proposals are carried out by expert engineers during dam reviews. With the perspective to assist these expert engineers, it is of great importance to develop methods and tools to manage the dynamic behaviour of dams and to model the processes at the same level of abstraction that is used by experts. In this chapter, the authors tackle the cognitive process of the diagnosis by means of a formal multi-modelling method and a diagnosis algorithm. The multi-modelling method called Timed Observations Method for Diagnosis (TOM4D) is based on the elaboration of four models: a Structural Model describing the relations between the components of the system, a Functional Model providing the relations between the values of the process variables (i.e. a set of mathematical functions), a Behavioural Model defining the states of the process and the discrete events firing the state transitions,

M. Le Goc (✉) · L. Torres
Laboratoire de Sciences de l'Information et des Systèmes, Aix-Marseille Université,
Avenue Escadrille Normandie-Niemen, 13397 Marseille Cedex 20, France
e-mail: marc.legoc@lsis.org

L. Torres
e-mail: lucile.torres@lsis.org

I. Fakhfakh · C. Curt
Irstea, 3275 route de Cézanne, CS 40061 13182 Aix-en-Provence Cedex 5, France
e-mail: ismail.fakhfakh@lsis.org

C. Curt
e-mail: corinne.curt@irstea.fr

© Springer-Verlag Berlin Heidelberg 2015
J.-M. Mercantini and C. Faucher (eds.), *Risk and Cognition*,
Intelligent Systems Reference Library 80, DOI 10.1007/978-3-662-45704-7_7

155

and a Perception Model composed of a set of abstract variables, a set of thresholds associated to these variables and a set of constraints. The resulting process allows the automatic fault detection, identification and diagnosis and it is applied to hydraulic dam safety.

7.1 Introduction

Technological systems must be designed and controlled to produce or provide a service while guaranteeing the safety of employees, users and consumers, of the environment and facilities. This safety can be jeopardized by the degradation of technological systems due to aging or design and construction defects, leading to dramatic events such as blasts, fires and the failure of hydraulic structures.

Controlling safety requires examining three types of process—physical (e.g. aging of civil engineering structures), informational (data collection and processing) and decisional (data aggregation for decision-making)—and taking into account different constraints, notably economic and regulatory ones. These 3 types of process cannot be considered as independent. Improving informational and decisional processes with the aim of controlling physical processes relies on the development of measurement, assessment and control methods, as well as diagnosis and prognostic models and algorithms. This chapter describes a diagnosis-based approach of risk assessment that is applied to the safety of a hydraulic dam.

Dams are heterogeneous structures characterized by complex behaviors that evolve through time due to natural aging. This aging can be accelerated by various causes: climatic conditions, poor design or construction conditions, events such as floods and earthquakes, insufficient or inadequate maintenance, etc. These causes lead to the development of more or less dependent deterioration phenomena. These phenomena contribute to the deterioration of dam reliability and safety through time and can finally lead to dam failure. Dams can then be sources of hazard for their environment and the population.

Moreover, the temporal aspect of the safety deterioration of a dam is of the main importance: except when provoked by external events such as earthquakes or floods, a failure is always preceded by gradual deteriorations of the dam components. It is then crucial to be able to identify the current state of a dam (supervision) and its past deteriorations (diagnosis), and to forecast its future evolution (prognosis) with the aim of proposing corrective actions and the associated time scales (control). Corrective actions are of various types: emergency actions, such as partial or complete emptying of the reservoir; major reconstruction, rehabilitation or safety projects; maintenance actions, such as drain outlet cleaning, scraping of the downstream slope, renewal of monitoring devices; upgrading dam safety monitoring by, for example, increasing measurement frequency, performing laboratory tests, etc. Controlling dam safety requires then on the one hand, to consider multiple and possibly interdependent phenomena and, on the other hand, to consider the temporal properties of these phenomena and theirs interdependencies. It is then particularly important to develop modeling and diagnosis approaches that take into account these two aspects and that allow to combine the raw data process with the cognitive models used by the experts to assess the safety of a dam.

Only very few approaches allows to reach these two requirements. Among these, the Timed Observation Theory [1] provides the mathematical and the methodological tools for the modeling, the supervision, the diagnosis and the prognosis of complex dynamic processes. In particular, the TOM4D methodology (Timed Observations Modeling For Diagnosis) [2–4] allows to build a quadripartite model of a dynamic process that can be used for an efficient diagnosis task. One of the main advantages of the TOM4D methodology is its ability to combine raw process data with expert's knowledge to provide the abstract (or conceptual) model of a process an expert uses to formulate its diagnosis knowledge [5]. TOM4D is then particularly well suited to provide an adequate model of a dam with the aim to ensure the compliance of its current state with the reliability and safety requirements, to diagnose the (past) causes of the deterioration of the reliability and the safety, and to forecast the reliability and safety evolutions at different time scales.

This chapter presents the application of the TOM4D methodology and its diagnosis algorithm to assess the compliance of a hydraulic dam with its reliability and safety requirements. To this aim, the next section recalls the main general requirements for the reliability and the safety of a hydraulic dam. Section 7.3 introduces the basis of the Timed Observations Theory and provides the main conceptual and mathematical tools that are required to assess a risk with the Timed Observation Theory (Sect. 7.4) and the introduction to the TOM4D methodology (Sect. 7.5). Section 7.6 presents the application of the TOM4D methodology to the French Sapins' dam modeling so that Sect. 7.7 can be focalized on the presentation of the diagnosis-based risk assessment with the TOM4D model of this dam. Finally, Sect. 7.8 concludes this chapter with a synthesis that draws some future perspectives of this work.

7.2 Dam Safety

Hundreds of thousands of dams are now in use throughout the world and some of them have been operating for several centuries.

Dams represent important economic stakes due to the numerous roles they fulfill: storing water for irrigation, producing hydroelectricity, supplying water to towns and businesses, etc. In 1997, the ICOLD (International Committee of Large Dams) counted 150,000 dams from 10 to 30 m high. In France, there are currently a total of 744 large scale dams and thousands of structures whose height is lower than 10 m. Worldwide storage capacity is about 6,000 km^3. Dams contribute to the management of the limited global water resource that is subject to poor distribution and considerable seasonal variations.

Dams are heterogeneous structures featured by complex behaviors that evolve through time because of their natural aging. This aging can be accelerated by environmental causes (climatic conditions, floods and earthquakes) or by internal causes (poor design or construction conditions, insufficient or inadequate maintenance...). These causes involve, during the life of the structure, the occurrence and the development of deterioration phenomena, more or less dependent and stemming from miscellaneous and complex sources. Such phenomena are for instance, fissuring of the concrete facing, clogging of the drain outlet, sliding of shoulders [6]. Figure 7.1 illustrates two examples of deterioration phenomena. The first photography (a) shows the deterioration of the protection of the upstream facing: cobblestones are lacking and the geotextile underneath is visible and locally deteriorated.

The second photography (b) shows the downstream shoulder of an embankment dam on which the vegetation is composed of shrubs and young trees. The death of this type of vegetation can generate specific water circulation caused by the disappearance of roots, leading to the deterioration of the sealing function.

These phenomena contribute to the deterioration of dam reliability and safety through time and can finally lead to dam failure. Consequently, dams can be sources of hazards for their environment and the surrounding population: annually, the average number of failures worldwide is from 1 to 2, and 160 cases of dam failures are well-documented. From 1959 to 1987, 30 dam failures were reported throughout the world, with the loss of 18,000 lives. In addition to casualties, both natural and economic environments are affected by dam failures, with possible domino effects if the wave caused by the failure reaches sensitive structures such as chemical or nuclear plants. In addition to these dramatic events, the deterioration of dam components due to accelerated aging can lead to economic losses caused by repairs, excessive water losses or the need to maintain the water at levels lower than normal reservoir level. Figure 7.2 presents an example of dam failure: the Malpasset Dam in France that failed in 1959, causing 421 victims and major material damage.

At present, all over the world, the assessment of dam reliability and safety and diagnoses are carried out by expert engineers at dam reviews where they make proposals for corrective actions. Deterioration and failure models and scenarios have been proposed in the literature.

Fig. 7.1 Examples of deterioration phenomena (photos Irstea)

Fig. 7.2 Failure of the Malpasset Dam (France) in 1959 (photo Irstea)

Nevertheless, these methods present some limits: they are qualitative, e.g., based on the FMEA (Failure Modes and Effects Analysis) approach [7], or they only consider the future of the dam [8], or they fail to take into account the whole set of available data, in particular visual data. This latter point is particularly important, notably for embankment dams [9]. Indeed, it is crucial to develop methods and tools for managing the dynamic behavior of dams and modeling the process at the same level of abstraction as that used by experts in their diagnoses.

7.3 Introduction to the Timed Observations Theory

The Timed Observations Theory (TOT) [1] provides a general framework for modeling dynamic processes from timed data by combining the Markov Chain Theory, the Poisson Process Theory, the Shannon's Communication Theory [10] and the Logical Theory of Diagnosis [11, 12].

This section aims at introducing the main concepts of the TOT, required in order to introduce the TOM4D KE methodology, that is to say the notions of observed process, timed observation, observation class and timed sequential binary relations between observation classes.

7.3.1 Observed Process

The Timed Observations Theory defines a dynamic process as an arbitrarily constituted set $X(t) = \{x_1(t), x_2(t), \ldots, x_n(t)\}$ of n functions $x_i(t)$ of continuous time $t \in \Re$. The set $X(t)$ of functions implicitly defines a set $X = \{x_1, x_2, \ldots, x_n\}$ of n variable names x_i.

According to the TOT, a dynamic process $X(t)$ is monitored by an abstract observer program denoted $\Theta(X, \Delta)$ which observes the functions $x_i(t)$ of $X(t)$ to describes their evolution over time with a finite set $\Delta = \{\delta_i\}$, $i = 1 \ldots m$, of constants δ_i (i.e. a number or a string). The program $\Theta(X, \Delta)$ usually accounts for the functions progression through timed messages recorded in a database. These messages can be alarms, warnings or reporting events. The program $\Theta(X, \Delta)$ is considered as an abstract observer because it can be implemented with a standard computer that records the timed messages in a database (i.e. a monitoring program) or these messages can be provided and recorded by a human, typically an expert, when analyzing the evolutions of a dynamic process.

Definition 7.1 Observed process Let $X(t) = \{x_i(t)\}$, $i = 1 \ldots n$, be a finite set of time functions; let $X = \{x_i\}$, $i = 1 \ldots n$, be the corresponding finite set of variable names; let $\Delta = \{\delta_j\}$, $j = 1 \ldots m$, be a finite set of constant values; let $\Theta(X, \Delta)$ be a program observing the evolution of the functions of $X(t)$.

The couple $(X(t), \Theta(X, \Delta))$ is an observed process.

The TOT considers a message at time t_k in a database as a timed observation (δ_i, t_k) where δ_i is a constant value of Δ and t_k is the moment at which the observation is considered to occur. For example, let us suppose that timed data recorded in a database are of the form "yymmddhhmmss, message" where yymmddhhmmss is a time stamp and message is a value determined by a monitoring program $\Theta(X, \Delta)$, "," being a separator character. According to the TOT, the message "080313132225, high" will be represented with a timed observation (δ_i, t_k) where $t_k = 080313132225$ and $\delta_i =$ high. That is to say, $(\delta_i, t_k) = ($high$, 080313132225)$.

In general terms, a timed observation (δ_i, t_k) is written by an abstract observer program $\Theta(X, \Delta)$ when a function $x_i(t)$ of continuous time enters in a specific interval of values. The specification of such an abstract observer program refers to a threshold value $\Psi_i \in \Re$ and two immediately successive values $x_i(t_{k-1}) \in \Re$ and $x_i(t_k) \in \Re$ so that:

$$x_i(t_{k-1}) < \Psi_i \wedge x_i(t_k) \geq \Psi_i \Rightarrow write((\delta_i, t_k)) \tag{7.1}$$

Generally speaking, in such a program, the action "$write((\delta_i, t_k))$" denote the action of recording a timed observation (δ_i, t_k) in a memory whenever is satisfied a particular predicate $\Theta(x, \delta, t)$ (here $x_i(t_{k-1}) < \Psi_i \wedge x_i(t_k) \geq \Psi_i$). For example, the Fig. 7.3 illustrates a function $x_i(t)$, where values above Ψ_i are interpreted by an observer program $\Theta(\{x_i\}, \{$high$\})$ as a high level of $x_i(t)$; that is, when

Fig. 7.3 Spatial
segmentation of the
$x_i(t)$ function of time

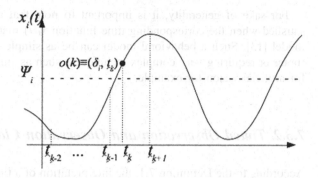

$x_i(t) \in [\Psi_i, +\infty)$. Thus, given a sequence of values $\omega = (x_i(t_1), x_i(t_2), \ldots, x_i (t_{k-1}), x_i(t_k), x_i(t_{k+1}))$, the program $\Theta(\{x_i\}, \{\text{high}\})$ will write a timed observation (high, t_k), which indicates that the function $x_i(t)$ entered the interval $[\Psi_i, +\infty)$ at time t_k. This principle is called the "spatial segmentation principle".

The Theory of Timed Observations establishes that the existence of a timed observation (δ_i, t_k), recorded in a database, allows to infer that the mentioned observation has been recorded by an unknown program $\Theta(\{x_i\}, \{\delta_i\})$ which implements the abstract logical equation described in (7.2).

$$\forall t_k \in \Gamma, \theta(x_i, \delta_i, t_k) \in \Theta \Rightarrow (\delta_i, t_k) \in \Omega \tag{7.2}$$

This equation associates the set Θ of all the assignations to a ternary predicate $\theta(x_\theta, \delta_\theta, t_\theta)$ with the set Ω of all the timed observations carried out by the program $\Theta(\{x_i\}, \{\delta_i\})$ (i.e., the database). A timed observation (δ_i, t_k) is then interpreted as the logical consequence of the assignation of the values x_i, δ_i and t_k to a ternary predicate $\theta(x_\theta, \delta_\theta, t_\theta)$. In other words, this means that the timed observation (δ_i, t_k) was recorded when the program $\Theta(\{x_i\}, \{\delta_i\})$ assigned the values x_i, δ_i and t_k to the predicate $\theta(x_\theta, \delta_\theta, t_\theta)$.

Given the sentences (1.1) and (1.2), the general meaning "is" can be always provided to the predicate θ so that the timed observation (δ_i, t_k) is interpreted as "at time t_k, x_i is δ_i". Considering that x_i is associated with a function $x_i(t)$, the meaning "equal" can also be attributed to the predicate θ, which leads to the following abuse of language: $\theta(x_i, \delta_i, t_k)$ means "Equal(x_i, δ_i, t_k)" (i.e. "$x_i(t_k) = \delta_i$"). Consequently, the Timed Observations Theory considers that a message contained in a database is a timed observation (δ_i, t_k) written by a program $\Theta(X, \Delta)$ which observes a time function $x_i(t)$ and implements the abstract Eq. (7.2). In our example, the timed observation (high, t_k) indicates that a program $\Theta(\{x_i\}, \{\delta_i\})$, observing a time function $x_i(t)$ and defining implicitly a predicate $\theta(x_\theta, \delta_\theta, t_\theta)$, has considered $\theta(x_i,$ high, $t_k)$ true and then it has written the timed observation (high, t_k) in the database Ω. This example illustrates the abuse of language frequently carried out, which associates the meaning "$x_i(t_k) = \text{high}$" with the interpretation of the level of the quantity "$x_i(t)$".

For sake of generality, it is important to note that a predicate $\theta(x_\theta, \delta_\theta, t_\theta)$ is satisfied when the corresponding time function $x_i(t)$ matches against a behavioral model [13]. Such a behavioral model can be as simple as the switch of an interrupter or requiring very complex techniques, such as signal processing techniques for artificial vision for example.

7.3.2 Timed Observation and Observation Class

According to the Definition 7.1, the interpretation of a timed observation (δ_i, t_k) is precisely the assigned predicate $\theta(x_i, \delta_i, t_k)$.

It is noteworthy that the program $\Theta(\{x_i\}, \{\delta_i\})$ could have errors: a timed observation (δ_i, t_k) could have been written in a database although the assertion $\theta(x_i, \delta_i, t_k)$ is not really true.

Definition 7.2 Timed Observation Let $X(t) = \{x_i(t)\}$, $i = 1 \ldots n$, be a finite set of time functions; let $X = \{x_i\}$, $i = 1 \ldots n$, be the corresponding finite set of variable names; let $\Delta = \{\delta_j\}$, $j = 1 \ldots m$, be a finite set of constant values; let $\Theta(X, \Delta)$ be a program observing the evolution of the functions of $X(t)$; let $\Gamma = \{t_k\}$, $t_k \in \mathfrak{R}$, $k \in N$ be a set of arbitrary time instants (i.e. a stochastic clock); and let $\theta(x_\theta, \delta_\theta, t_\theta)$ be a predicate implicitly determined by $\Theta(X, \Delta)$. Then,

- a *timed observation* $(\delta_j, t_k) \in \Delta \times \Gamma$ on $x_i(t)$ is the assignation of values x_i, δ_j and t_k to the predicate $\theta(x_\theta, \delta_\theta, t_\theta)$ such that $\theta(x_i, \delta_j, t_k)$;
- by definition $o(t_k)$ denotes a timed observation (i.e., $o(t_k)$ α (δ_j, t_k)) and,
- a scenario ω_i is an ordered sequence of r timed observations; that is, $\omega_i :$ $\{1, \ldots, r\} \rightarrow \Delta \times \Gamma \mid r \in N \wedge \forall i, j \in \{1, \ldots, r\}, i < j, \omega(i) = o(t_k) \wedge \omega(j) = o(t_r) \Rightarrow t_k \leq t_r\}$;

We denote $\Omega = \{\omega_i\}$, $i = 1 \ldots l$, the set of all the scenario ω_i so that ($*\omega_i \in \Omega$) $\Delta \times \Gamma$.

Moreover, timed observations $o(t_k) \equiv (\delta j, t_k)$ on a particular function $x_i(t)$ implicitly determine a variable x_i, which assumes discrete values $\delta_{j1}, \delta_{j2}, \ldots$, and describes the evolution of $x_i(t)$ according to the interpretation implemented in the observer program. That is to say, when $\Theta(X, \Delta)$ considers $\theta(x_i, \text{high}, t_k)$ true and then writes (high, t_k) in Ω, it is implicitly defining a discrete variable x_i which assumes at least the value "high".

Consequently, a timed observation $o(t_k)$ α (δ_j, t_k) and the implicit existence of an associated discrete variable x_i enable to define the notion of observation class, other important concept of the Timed Observations Theory. An observation class is a set $C_{xi} = \{(x_i, \delta_j) \mid \delta_j \in \Delta\}$ that associated a variable x_i with the constant values δ_j it assumes. Thus, this concept establishes the link between a constant $\delta_j \in \Delta$ and a variable $x_i \in X$ and then, a timed observation $o(t_k)$ α (δ_j, t_k) is an occurrence of an observation class $C_{xi} = \{(x_i, \delta_j) \mid \delta_j \in \Delta\}$. Definition 7.3 specifies this concept.

Definition 7.3 Observation Class Let $X(t) = \{x_i(t)\}$, $i = 1...n$, be a finite set of time functions; let $X = \{x_i\}$, $i = 1...n$, be the corresponding finite set of variable names; let $\Delta = \{\delta_j\}$, $j = 1...m$, be a finite set of constant values; let $\Theta(X, \Delta)$ be a program observing the evolution of the functions of $X(t)$; and let $\Delta = \cup \Delta x_i$ be such that $\Delta x_i = \{\delta_j\}$ is a set of constant values which can be assumed by x_i, $\forall x_i \in X$

An observation class C_i associated with a variable $x_i \in X$ is a set $C_i = \{(x_i, \delta_j) \mid \delta_j \in \Delta x_i\}$.

One of the major consequences of the notion of observation class is a decompositional principle that is derived from the superposition theorem of the TOT. This decompositional principle can be formulated as follow:

- Given an observed process $(X(t), \Theta(X, \Delta))$ where the program $\Theta(X, \Delta)$ is memoryless and choose the constants δ^i of Δ independently, defining a set $C = \{C_i\}$ of m observation classes $C_i = \{(x_i, \delta_j)\}$ corresponds to the decomposition of $(X(t), \Theta(X, \Delta))$ in a set of m observed sub-processes $(X_i(t), \Theta_i(X_i, \Delta_i))$ so that $(X(t), \Theta(X, \Delta)) = \cup_i(X_i(t), \Theta_i(X_i, \Delta_i))$.

The memoryless and the independence conditions of the program $\Theta(X, \Delta)$ are, in practice, very easy to check. This decompositional principle comes from a fundamental theorem of the Timed Observations Theory called the "superposition theorem" [1]. The important consequence is that the m observed sub-processes $(X_i(t), \Theta_i(X_i, \Delta_i))$ defines, by construction, m stochastic clocks $\Gamma_i = \{t_{ki}\}$, $t_{ki} \in \Re$, $k_i \in N$ so that $\Gamma = *\Gamma_i$.

The fundamental interest of this property resides in the definition of an observation class as a singleton $C_i = \{(x_i, \delta_j)\}$, $\delta_j \in \Delta x_i$ (i.e. C_i contains one and only one couple $C_i = \{(x_i, \delta_j)\}$). In that case, the m observation classes of C decomposes an observed process $(X(t), \Theta(X, \Delta))$ in m minimal processes $(\{x_i(t)\}, \Theta_i(\{x_i\}, \{\delta_j\}))$ that produces m stochastic clocks $\Gamma_i = \{t_{ki}\}$, $t_{ki} \in \Re$, $k_i \in N$, each of them marking the time t_{ki} of the assignation "$x_i(t_{ki}) = \delta_i$". This facilitates the analysis of the relations between the functions $f_i(t)$ making the core of the process $X(t)$ so that a structure can be provided to $X(t)$.

7.3.3 Binary Temporal Relations

A basic concept of the Timed Observations Theory is the concept of binary temporal relation:

Definition 7.4 Temporal binary relation A temporal binary relation $r(C_i, C_j, [\tau^-, \tau^+])$ is an oriented (sequential) relation between two observation classes $C_i = \{(x_i, \delta_i)\}$ and $C_j = \{(x_j, \delta_j)\}$ that is timed constrained with the $[\tau^-, \tau^+]$ interval $(\tau^-, \tau^+ \in \Re)$.

The temporal constraint $[\tau^-, \tau^+]$ of a temporal binary relation $r(C_i, C_j, [\tau^-, \tau^+])$ is the time interval for observing a timed observation $C_j(t_k) \equiv (\delta_j, t_k)$ of the "output" observation class C_j after the observation of a timed observation $C_i(t_{k-n}) \equiv (\delta_i, t_{k-n})$ of the "input" observation class C_i:

Definition 7.5 Observing a temporal binary relation Let the couple $(X(t), \Theta(X, \Delta))$ be an observed process defining a particular set $C = \{C_i\}$ of m observation classes containing two classes $C_i = \{(x_i, \delta_j)\}$ and $C_j = \{(x_j, \delta_j)\}$. Let $\omega = \{\dots, C_l(t_k), \dots\}$, $t_k \in \Gamma \subseteq \Re$, $k = 0 \dots n - 1$, $l = 0 \dots m - 1$, be a sequence of n timed observations $C_l(t_k)$ provided by $(X(t), \Theta(X, \Delta))$.

A timed binary sequential relations $r_{ij}(C_i, C_j, [\tau^-, \tau^+])$ between the classes C_i and C_j is said to be observed in ω if there is at least two timed observations $C_i(t_k - n)$ and $C_j(t_k)$ so that $t_k - t_{k-n}$ satisfies the timed constraint $[\tau^-, \tau^+]$ of $r_{ij}(C_i, C_j, [\tau^-, \tau^+])$.

Formally, the relation $r_{ij}(C_i, C_j, [\tau^-, \tau^+])$ is observed if and only if:

$$r_{ij}(C_i, C_j, [\tau^-, \tau^+]) \Leftrightarrow \exists C_i(t_{k-n}), C_j(t_k) \in \omega \wedge t_k - t_{k-n} \in [\tau^-, \tau^+] \qquad (7.3)$$

Naturally, the confidence with the representativity of a temporal binary relation $r_{ij}(C_i, C_j, [\tau^-, \tau^+])$ is linked with the ratio of the number of couples $(C_i(t_{k-n}), C_j(t_k))$ in ω with the total number of couples $(C_a(t_a), C_b(t_b))$, $t_a - t_b \in [\tau^-, \tau^+]$ that ω allows to build. This explains the introduction of the notion of probabilities in the TOT framework that leads to a new Knowledge Discovery in Database process called TOM4L (Timed Observation Mining for Learning, [14–17]) the presentation of which is out of the scope of this chapter. Whatever is the way of building a set of temporal binary relation $r_{ij}(C_i, C_j, [\tau^-, \tau^+])$, it constitutes a model:

Definition 7.6 Abstract chronicle model An abstract chronicle model is an arbitrarily made set $M = \{\dots, r_{ij}(C_i, C_j, [\tau_{ij}^-, \tau_{ij}^+]), \dots\}$ of temporal binary relations of the form $r_{ij}(C_i, C_j, [\tau_{ij}^-, \tau_{ij}^+])$.

The abstract chronicle models of the TOT framework are represented with a graphical knowledge representation language called "ELP" for "Event Language for Process behavior modeling" [17]. An instance ω_i of an Elp Model M is a sequence of timed observations that is consistent with the logical and the timed constraints of M.

For example, let us consider the following abstract chronicle model $M_{123} = \{r_{12}(C_1, C_2, [0, 5]), r_{23}(C_2, C_3, [3, 8])\}$; the sequence $\{C_1(1), C_4(3), C_2(4), C_1(8), C_3(10)\}$ contains the occurrences $C_1(1)$, $C_2(4)$ and $C_3(10)$ satisfy the logical and the timed constraints of M_{123}:

- $C_1(1)$ and $C_2(4)$ satisfy the logical condition of the relation $r_{12}(C_1, C_2, [0, 5])$ (i.e. the observation class of $C_1(1)$ is C_1 (resp. C_2 for $C_2(4)$).
- $C_1(1)$ and $C_2(4)$ satisfy the temporal condition of the relation $r_{12}(C_1, C_2, [0, 5])$ (i.e.4 − 1 = 3, 3 ∈ [0, 5]).
- $C_2(4)$ and $C_3(10)$ satisfy the logical condition of the relation $r_{23}(C_2, C_3, [3, 8])$ (i.e. the observation class of $C_2(4)$ is C_2 (resp. C_3 for $C_3(10)$).
- $C_2(4)$ and $C_3(10)$ satisfy the temporal condition of the relation $r_{23}(C_2, C_3, [3, 8])$ (i.e. 10 − 4 = 6, 6 ∈ [3, 8]).

The notion of abstract chronicle model is of the most interest for the diagnosis based risk assessment. In particular, the abstract chronicle models that constitute a path:

Definition 7.7 Path An abstract chronicle model made with a suite of n − 1 timed sequential binary relations $P_i = \{ r(C^i, C^{i+1}, [\tau_1^-, \tau_1^+]), r(C^{i+1}, C^{i+1}, [\tau_2^-, \tau_2^+]), \ldots, r(C^{n-1}, C^n, [\tau_n^-, \tau_n^+]) \}$ is a path.

The M_{123} model, for example, is a path. The notion of path provides a meaning to a sequence ω that satisfies all the relations of a path. This meaning is a kind of history of the observed process $(X(t), \Theta(X, \Delta))$ that has produced ω. This meaning is provided by the suite of temporal binary relation linking together a set of observation classes.

This introduction of the main concepts of the Timed Observations Theory allows to provide a formulation of the TOT fundamental theorem of induction:

Definition 7.8 Induction from timed observations Let the couple $(X(t), \Theta(X, \Delta))$ be an observed process defining a particular set $C = \{C_i\}$ of m observation classes containing two classes $C_i = \{(x_i, \delta_j)\}$ and $C_j = \{(x_j, \delta_j)\}$. Let ω a sequence of timed observation provided by $(X(t), \Theta(X, \Delta))$. Let $\omega_i = \{\ldots, C_i(t_{ki}), \ldots\}$, $t_{ki} \in \Gamma_i \subset \Gamma \subseteq \Re, k_i = 0 \ldots n_i - 1$, and $\omega_j = \{\ldots, C_j(t_{kj}), \ldots\}$, $t_{kj} \in \Gamma_j \subseteq \Re, k_j = 0 \ldots n_j - 1$, be respectively two sequences of n_i and n_j timed observations of the classes C_i and C_j provided by $(X(t), \Theta(X, \Delta))$ (i.e. $C_i \in C$ and $C_j \in C$, $\omega_i \subset \omega$, $\omega_j \subset \omega$ and $\omega_i \cap \omega_j = \Phi$).

Inducing a binary relation $r_\alpha(\Gamma_i, \Gamma_j)$ between two stochastic clocks Γ_i and Γ_j from two sequences ω_i and ω_j corresponds to induce a binary temporal relations of the form $r_{ij}(C_i, C_j, [\tau_{ij}^-, \tau_{ij}^+])$ which subsumes the existence of a relation between the variables x_i and x_j respectively.

The importance of the concept of observation class in the TOT framework clearly appears with this theorem: assuming that there is a relation between two observations classes C_i and C_j corresponds to assuming that there is a relation between the corresponding variables x_i and x_j, that is to say, by construction, between the two corresponding functions $x_i(t)$ and $x_j(t)$ of the dynamic process $X(t)$. This theorem establishes the relations between the constants δ_i, the time stamps Γ_i and the functions $f_i(t)$ that are organized over the notion of variable as Fig. 7.4 aims at illustrating.

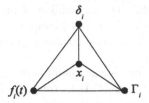

Fig. 7.4 Relations between the basic objects of the TOT

In other words, defining a variable x_i means defining six binary relations linking together the variable x_i, a function $f_i(t)$, a constant δ_i and a stochastic clock Γ_i. This provides a powerful tool to analyze the logical coherence within a knowledge corpus about a dynamic process:

Definition 7.9 Coherence law What is assumed about the relations between two variables x_i and x_j must be coherent with what is assumed about the relations between the corresponding constants, observation classes, stochastic clocks and so, the corresponding functions $x_i(t)$ and $x_j(t)$, and reversely.

This explains why this theorem is a fundamental basis of the TOM4D KE methodology. The next section presents the principles of the TOM4D methodology that are grounded on this theorem.

7.4 Risk Assessment with the TOT

Generally speaking, risk assessment is a knowledge intensive task that requires a model of the process under consideration.

According to the Timed Observations Theory, the notion of risk is concerned with particular behaviors of a dynamic process: a risk is linked with the occurrence of an undesired behavior of an observed process $(X(t), \Theta(X, \Delta))$. This means that the required knowledge to assess a risk about a dynamic process is the one required by the tasks of monitoring, diagnosis and prognosis:

- the monitoring task requires the knowledge to infer the current behavior and to categorize it as desirable or undesirable behavior,
- the diagnosis task requires the knowledge to infer the causes of the undesirable behaviors,
- and the prognosis tasks requires the knowledge to infer the potential future undesirable behaviors that can result from the current situation described by the monitoring and the diagnosis tasks.

The quality of the knowledge corpus required by these tasks is directly linked with the pertinence of the risk assessment task. If this point is quite trivial, the knowledge engineering method used to acquire and represent this knowledge corpus must provide the tools to guarantee an adequate level of quality. As it has been stated earlier, the basic concepts of the Timed Observations Theory provide the way to build such tools.

7.4.1 Knowledge

Generally speaking, knowledge results of the interaction between an information flow and an arbitrary purpose.

This interaction is assumed by humans, which define their purpose according to their own expectations [18–20]. Information comes from all the possible sources: believes, observations, experimentations, scientific axioms, sensors, etc. [21–23]. The interaction is basically an interpretation of the information flow that traverses a thinking human [24, 25]. To build the TOM4D KE methodology, we uses the following operational definition of knowledge:

Definition 7.10—Operational notion of knowledge Knowledge results from an intentional interpretation of a flow of information.

This definition establishes a relation between knowledge, information and a purpose (an intention). The purpose is always defined by humans: in the framework of the TOT, the purpose is implemented in an observer program $\Theta(X, \Delta)$ which can be either "executed" by a human or a computer. As the hydraulic dam example will show, the purpose is typically the assessment of a risk linked with the occurrence of an undesired behavior.

According to the Timed Observations Theory, the information flow is a sequence $\omega = \{\ldots, C_i(t_k), \ldots\}$ of timed observations $C_i(t_k) \equiv (\delta_i, t_k)$. By construction, ω defines a set $C = \{C_i\}$ of observation classes $C_i = \{(x_i, \delta_i)\}$. This set provides the reading keys allowing to provide the interpretation of ω as a sequence $\Theta(\omega) = \{\ldots, \theta(x_i, \delta_i, t_k), \ldots\}$ of assignations $\theta(x_i, \delta_i, t_k)$. The C set must then be constituted according to a particular purpose: this means that each observation class C_i of C contains a piece of a global purpose so that a path (cf. Definition 7.7) represents a model of scenarios according to this global purpose. A path being a suite of temporal binary relation of the form $r_{ij}\left(C_i, C_j, \left[\tau_{ij}^-, \tau_{ij}^+\right]\right)$ and each observation classes C_i and C_j being made accordingly to a particular purpose, the Definition 7.10 of knowledge can then be re-formulated according to the TOT fundamental theorem of induction (Definition 7.8):

Definition 7.11—Knowledge according to the TOT Any relation that is logically consistent with a binary temporal relation of the form $r_{ij}(C_i, C_j, [\tau_{ij}^-, \tau_{ij}^+])$ is a piece of knowledge.

As a consequence, considering the assessment of a risk concerning an observed process $(X(t), \Theta(X, \Delta))$, the aim of the knowledge engineering phase is the elicitation of an adequate knowledge corpus and its representation under the form of a model in order to make possible the emergence of specific patterns that are directly linked with a risk to assess. The definition of a risk is then crucial for modeling a dynamic process: in particular, it specifies de facto the abstraction level of the model.

7.4.2 Model

The fundamental role of a model is the sharing of knowledge between humans.

This sharing is facilitated through the mediation of signs belonging to a particular set (often called "alphabet"). These signs have no meaning in themselves but are necessary to represent knowledge in order to share a common understanding of an observed set of phenomenon.

As a consequence, a model is made with a particular arrangement of signs: the meaning results precisely of the specific arrangement the modeler choose to share its knowledge. The representation of a knowledge corpus requires then a set of rules that defines the authorized arrangements (i.e. the "grammar"). This leads to define the notion of model according to the TOT:

Definition 7.12—Model according to the TOT A model is an organized set of knowledge representations.

It is clear that the organization of the knowledge representations within a model is of the main importance. The Timed Observations Theory being concerned with the evolutions of a process over time, the knowledge under consideration is linked with the relations between function of time $f_i(t)$, constants δ_i and stochastic clocks $\Gamma_i = \{t_{ki}\}$, $t_{ki} \in \Re$, $k_i \in N$. The Timed Observations Theory organizes these relations around the notion of variable x_i (cf. Fig. 7.4). A piece of knowledge belongs then to three fundamental categories:

1. Relations between the functions $f_i(t)$ of the process $X(t)$.
   This category of knowledge is called the "structural knowledge" because the functions are the constituents (i.e. the "components") of the process $X(t)$.
2. Relations between the constants δ_i of the observer program $\Theta(X, \Delta)$.
   This category is called the "functional knowledge" because the relations between the constants δ_i of Δ can be represented with logical rules linking together sub-sets Δ_i of constants of Δ and so, specifies abstract mathematical functions under the form of "tables of values".
3. Relations between the stochastic clocks Γ_i.
   This category is called the "behavioral knowledge" because these relations describes the links between the evolutions of the functions $f_i(t)$ of the process $X(t)$.

As Fig. 7.4 aims at showing, these three categories of relations are linked together: a specific set Δ of constant δ_i will lead an observer program $\Theta(X, \Delta)$ to describe the evolution of a process $X(t)$ with a particular set Γ of time stamps. The role of the concept of "variable" in the Timed Observations Theory framework is to provide the mean to analyze the consistency of these three categories of knowledge about a process $X(t)$.

So, the concept of variable defines a supplementary category of knowledge, which is a kind on "meta-knowledge", that fundamentally defines the way a dynamic process $X(t)$ is perceived by humans (i.e. the modeling point of view). In practice, this leads to the following definition of the aim of the modeling activity:

Definition 7.13—Modeling activity according to the TOT The modeling activity of a dynamic process $X(t)$ aims at representing the elicited knowledge according to a formalism and at distributing the knowledge representations over three models, the structural, the functional and the behavioral model, according to a definition of a particular set of variables X.

By construction, the particular set of variables X is a subset of all the variables that can be defined about a dynamic process. The only rational way to specify X is precisely the modeling purpose: are only required the variables that play a role in the modeling purpose (i.e. a risk assessment). The other variables can be forgotten (at least in a first step). This set of variables X defines then the process according to a modeling purpose, and so fixes the abstraction level of the model.

Finally, a Knowledge Engineering methodology must defines the organization laws and the representation rules of knowledge (i.e. the representation formalism). Within the TOT framework, the formalism must allow the expression of relations between the main concept of the TOD. Recalling the Definition 7.10 of knowledge according to the TOT, a knowledge corpus concerning a given purpose about the behavior of an observed process $(X(t), \Theta(X, \Delta))$, typically the assessment of a risk, will be represented with an organized set of binary relation between functions $f_i(t)$, constants δ_i, stochastic clocks Γ_i and variables x_i (Fig. 7.4).

From this analysis, a set of knowledge modeling principles have been defined that are the basis of the TOM4D KE methodology.

7.5 TOM4D KE Methodology

TOM4D (Timed Observations Modeling for Diagnosis) is a knowledge engineering modeling approach for dynamic systems focused on timed observations. The aim of TOM4D is to produce suitable models for knowledge intensive tasks about dynamic process as for example supervision, diagnosis, prognostic and control tasks. Such knowledge intensive tasks constitute the core of any risk assessment function.

One of the main specificities of the TOM4D methodology is to allow the modeling from both timed observations and experts' a priori knowledge. In other words, a TOM4D process model made from experts' a priori knowledge presents the following major properties:

1. It can be faced with real world timed data (i.e. sequences of timed data making scenarios).
2. Any experts' assertion about the dynamic process can be faced with the TOM4D model.

These two major properties are fundamental when reasoning about a dynamic process: the first one allows to assess the adequacy of a model with real world data and the second allows the check the experts' knowledge with an already validated model. Furthermore, the second property constitutes a way to analyze the completeness and the coherence of a model under construction.

These two properties comes from the modeling principles that have been defined to design the TOM4D methodology.

7.5.1 TOM4D Modeling Principles

The aim of the TOM4D methodology is to allow the building of a model that (i) resides at the same abstraction level as the experts' knowledge and (ii) is logically coherent (i.e. contains no contradiction) and (iii) is as complete as possible. These goals are given in the order of their importance: clearly, providing a coherent model at the right abstraction level is the main modeling law of TOM4D, its completeness being desired but does not constitute a primary condition.

To assess the abstraction level of the model, TOM4D resorts to a knowledge interpretation framework in order to introduce, in the modeling process, the semantic content provided by experts in a gradual and controlled way. This knowledge interpretation framework is based on the conceptual models of the CommonKADS [26, 27] methodology to interpret the experts' knowledge, and uses the Tetrahedron of States (ToS) [28, 29] to analyze its soundness. This interpretation leads to define a particular set X of variables constituting the "core" of the modeled process so that the six types of binary relations of Fig. 7.4 can (must) be examined.

To assess the coherence of the model, the knowledge representations are distributed over the three basic models of the TOT, the structural, the functional and the behavioral models, according to the meaning of the corresponding knowledge. TOM4D has then been design as a primarily syntax-driven approach [2–4] so that Reiter's Logical Theory of Diagnosis [11] can be used to analyze the properties of the produced models according to the "Coherence law" of Definition 7.9, and in return, supply tools to assess the experts' knowledge [12]. To this aim, the knowledge representation formalism is based on the Formal Logic (i.e. the predicate calculus).

Keeping in mind that a TOM4D model aims at being the experts' model of a dynamic process, the experts' knowledge can be view as a set of propositions that the experts formulate according to this model. In other words, any proposition formulated by an expert is, by hypothesis, an assertion about a property of the process. These propositions are concerned with the structure, the functions, the behavior or the role the process play in a exploitation purpose. This means that the completeness of a TOM4D model corresponds to the property of such a model to allow to provide any proposition an expert can formulate about the corresponding process. So to assess the completeness of a TOM4D model, it necessary but sufficient to check first the completeness of the variable set X and next, to verify that all the possible binary relation between two variables x_i and x_j of X has been examined according to the three fundamental dimensions of knowledge that are the structural, the functional and the behavioral dimensions. Again, the syntactic based modeling approach of TOM4D facilitates the analysis of the completeness of a model.

As a consequence, the combination of a syntactic and a semantic approach in the TOM4D modeling principles allows to provide tools for the knowledge engineers to control the knowledge acquisition process and identify the main modeling concepts of the dynamic process (variables, constants, values, thresholds, components, states, etc.).

Clearly, the five fundamental modeling principles of TOM4D comes from the notion of variable of the Timed Observations Theory [1]:

1. Variable localization.

   A function $f_i(t)$ is a signal provided by a sensor located at a particular place defined as a component. So a function $f_i(t)$ specifies a variable x_i, a component c_i and a binary relation that associate x_i to c_i. As a consequence, a variable x_i is always associated with a sensor that is physically located on a component c_i. In other words, any variable x_i of X must be associated with one and only one component c_i.

2. Multi-value variable.

   A variable x_i is necessarily defined over a set Δ_{xi} of possible values containing at least two elements. This means that when the experts' knowledge defines only one value δ_i for a variable x_i, the knowledge engineer must introduce in Δx_i another constant, denoted δ_j for example, meaning "not δ_i" (i.e. $\Delta_{xi} = \{\delta_i, \delta_j\}$ and $\delta_j \equiv \neg \delta_i$). This principle is a direct consequence of the spatial segmentation of the Timed Observation Theory (cf. Fig. 7.4).

3. Discernible state.

   The assignment of a value δ_i to a variable x_i results necessarily of an observable modification in the dynamic process $P(t)$. Such an observable modification marks a state transition in $P(t)$. In other words, a state transition is conditioned by an occurrence $C_i(k) \equiv (\delta_i, t_k)$ of an observation class $C_i = \{(x_i, \delta_i)\}$, and a temporal binary relation $r_{ij}(C_i, C_j, \left[\tau_{ij}^-, \tau_{ij}^+\right)$ defines a unique "discernible state". Because the notion of "discernible state" is linked with the observer program $\Theta(X, \Delta)$, this notion is weaker than the classical "state" notion of the Discrete Event System (DES) domain.

4. Knowledge of different nature separation.

Since the TOT defines four categories of knowledge (structural, functional, behavioral and perception), four models will contains a specific category of knowledge representations: a Structural Model *SM* will contain all the structural knowledge, a Functional Model *FM* will contain all the functional knowledge, a Behavioral Model *BM* will contain all the behavioral knowledge, and a Perception Model *PM* will contain the perception knowledge (Fig. 7.5). This constitutes the multi-modeling framework of TOM4D [28, 30, 31].

5. Symbol driven modeling.

The knowledge interpretation aims at identifying the minimal set of symbols denoting a variable, a constant or a components and the minimal set of relations between them. The logical properties coming from these minimal sets are necessary and sufficient to complete the model. Among other meanings, this principle means that the introduction of a symbol that is not associated with an element of the domain knowledge is prohibited. In particular, a domain schema must be used to identify the concepts that play the role of variable in the knowledge corpus.

Figure 7.5 illustrates the conceptual framework of the modeling process of TOM4D. This conceptual framework is centered on the TOT notion of variable: the structural model *SM* associates any variables used in a function of the functional model *FM* with a component (or a component aggregate) and a timed observation class of the behavioral model *BM* associates a variable to a constant. So, a timed observation being an occurrence of an observation class, it corresponds to the assignment of a constant to a variable that play a role in at least one function of the functional model *FM* and is located with one of the components of the structural model *SM*. The perception model *PM* defines the minimal set of variables and the corresponding minimal set of constants that are required to define the goals of a dynamic process. Doing so, the perception model specifies a minimal set of constraints that the functional, structural and behavioral models must respects. This means that any relation defined in these models must be consistent with the constraints of the perception model *PM*.

These five principles constitutes a strong logical basis for the modeling work of the knowledge engineer: from the identification of a variable, the knowledge engineer will identify the possible values the variable can take over time, its

Fig. 7.5 Relations between TOM4D models

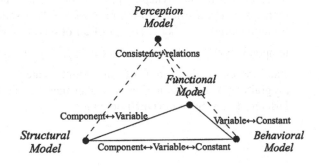

corresponding component and its observation classes. Doing so for all the variables, the knowledge engineer defines the state space of the dynamic process. Next, the knowledge analysis examines all the possible and the impossible relations between two elements (variable, constant, component and observation classes), conducted through their semantic properties. The organization of the resulting knowledge representations in the four models leads to an operational model of the dynamic system.

7.5.2 TOM4D Modeling Process

The modeling process aims to produce a generic model of a dynamic process from the available knowledge and data. The available knowledge contains, by definition, a description of the modeling goal: the assessment of some risks linked with the dynamic process exploitation.

The TOM4D modeling process is made with three main phases: knowledge interpretation, process definition and generic modeling. The logical dependencies of these three phases are represented in the conceptual graph of Fig. 7.6 where the

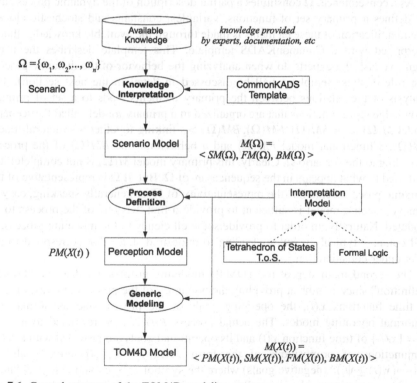

Fig. 7.6 General structure of the TOM4D modeling process

TOM4D modeling process is described as a reasoning process that provides a TOM4D model of a dynamic process given the available knowledge (i.e. a CommonKADS inference structure where ovals represents inference steps and rectangles defines the concepts' roles). Therefore, the exploitation method of such an inference structure must be defined according to the modeling problem: this explains why TOM4D is not a method but a methodology. How the control flow of the modeling process is carried out, is not part of this structure. Clearly, when a problem requires a modeling phase, any multi-modeling method is cyclical and each step can require to return back to previous phases with the objective of revising the expert's knowledge or modeling decisions.

The objective of the "Knowledge Interpretation" step is to define a scenario model given all the available knowledge, a set Ω of sequences ω_i of variable values over time and a CommonKADS template of the cognitive task (diagnosis template for example). The TOM4D methodology being firstly concerned with dynamic process, the temporal behavior of the process is of the main importance in the analysis. Because this behavior is, generally speaking, poorly described by the available knowledge, the scenario Ω is a fundamental input to the modeling process. Each sequence ω_i of Ω is an ordered set of measure of the values some process variables takes over time. These suites of timed measures describes a particular evolution of the process.

As a consequence, Ω constitutes a partial description of the dynamic process and so, defines a primary set of functions, variables, constants and stochastic clocks. The identification of these elements is made through the available knowledge that is interpreted with a CommonKADS template. This template describes the usual cognitive task the experts do when analyzing the behavior of the dynamic process (the role of these templates will be discussed in detail in the next section). The analysis of the relations between the primary elements leads to a set of primary knowledge representations that are organized in a primary model called the scenario model $M(\Omega) = <SM(\Omega), FM(\Omega), BM(\Omega)>$ linking together a structural model $SM(\Omega)$, a functional model $FM(\Omega)$ and a behavioral one $BM(\Omega)$ of the process according to the scenario Ω. Clearly, this primary model $M(\Omega)$ is not complete: it is restricted to what appears in the sequences ω_i of Ω. But, if Ω is representative of the dynamic process behavior (the representativity of Ω is, generally speaking, easy to achieve), such a model is sufficient to provide a first "vision" of the process to be modeled. Naturally, in order to provide some efficiency to the modeling process, Ω will contains typical evolutions leading to undesired states that correspond to the risks under consideration.

The second main step of the TOM4D modeling process is called the "Process Definition" since it aims at providing the boundary of the process in terms of a set of time functions $x_i(t)$, the operating goals of the process and its normal and abnormal operating modes. The actual process $P(t)$ is then restricted to a set $X(t) = \{x_i(t)\}$ of time function $x_i(t)$ and its operational goals are provided with a set of conjunctions of propositions of the form "$\forall t, x_i(t) \geq \psi_i$" (positive goals) or "$\forall t, \neg(x_i(t) \geq \psi_i)$" (negative goals) where the symbol "$\neg$" denote the logical "not". Clearly, a negative goal is typically a direct formulation of a risk linked with the exploitation of the process $P(t)$. The operating modes are also represented with a set of conjunctions of the same form. This set is partitioned in two sets, the set of the desired and the undesired modes respectively called the "normal behavior set" and the "abnormal behavior set" according to Reiter's theory of Diagnosis [11]. The input knowledge of the "Process Definition" step are the scenario model $M(\Omega)$ resulting of the "Knowledge Interpretation" step and the conceptual frameworks of the Formal Logic and the Tetrahedron of States (ToS) [29] of the Newton classical physics. These two frameworks constitutes the only semantic contexts allowing the logical and the physical interpretation of the modeling symbols used to denote the variables and the constants defined in the scenario model $M(\Omega)$. The role of these frameworks is to provide the set of laws allowing the knowledge engineer to control the representation of the semantic in the model and consequently the interpretation of the binary relations of the model. This phase being an important and delicate point in the TOM4D methodology, it will be discussed with more details in the next section. The output of the "Process Definition" step is the "Perception Model" PM $(X(t))$ of the dynamic process since it defines the way the process is perceived by the experts: nothing but what can be derived from this model can be taken into account the structural, the functional and the behavioral model of the process. In other words, $PM(X(t))$ defines the level of abstraction the expert use to reason about the process $P(t)$.

The last step, the "Generic Modeling", defines the set $X = \{x_i\}$ of the variables with their definition domain $\Delta_{xi} = \{\delta_i\}$, identifies the corresponding sets of components, observation classes and logical relations between the constants of the definition domain of the variable, and distributes the representation of the pertinent binary relations over the three models, that is to say the structural model $SM(X(t))$, the functional model $FM(X(t))$ and the behavioral model $BM(X(t))$. The objective is then to define a model $M(X(t)) = <PM(X(t)), SM(X(t)), FM(X(t)), BM(X(t))>$ of a type of process that is coherent with $M(\Omega)$, the scenario model, but that generalizes it: this is the meaning of the usage of the "generic" attribute to qualify the generality and the abstraction level of the resulting model $M(X(t))$.

The "Generic Modeling" step is accomplished using the Perception Model $PM(X(t))$ and the available knowledge, according to the representation and the interpretation laws defined according to the Formal Logic and the ToS frameworks. These frameworks allow the systematic exploration of the whole semantic and syntactic spaces that constitutes the global modeling space: the semantic space is defined with the physical dimension of the variables x_i (typically according to the International System of Units), and the syntactic space is defined as the matrix of all the pairs (a, b) that can be made with the alphabet of the symbols used to represents the knowledge.

The next section discusses the way a knowledge engineer can control the introduction of semantic elements in the model with the use the Formal Logic and the ToS frameworks and the TOM4D formalism will be presented through the application of the TOM4D modeling process to a real world example, the Sapins' dam in France.

7.5.3 Controlling the Semantic

One of the main difficulty with the application of Knowledge Engineering is the analysis of the semantics contained in a knowledge corpus provided by an expert. By definition, the knowledge engineer is a "novice" compared with the experts of the domain and consequently, the knowledge engineer has not the necessary distance to analyze the coherence and the scope of a new piece of knowledge. This difficulty increases drastically when working with a dynamic process: it is very easy for a "novice" to admit propositions that seems physically reasonable but are not coherent with the current version of the knowledge model. Knowledge Engineering researches aims at providing tools to facilitate the semantic analysis.

In particular, CommonKADS [26, 27] is a Knowledge Base System (KBS) engineering methodology which offers a structured approach in the management of the development of KBSs. CommonKADS is well known for the fundamental role that this methodology attributes to the "Conceptual Model of Expertise" (CME) in the development process of a KBS. This model describes the types and structures of the knowledge required to accomplish a particular cognitive task and thus, it acts as a tool that helps to clarify the structure of a knowledge-intensive information-processing task. This model is developed, in a way that is understandable by humans, as part of the analysis process and therefore, it does not contain any implementation-specific term. Thus, this one is an important vehicle for communication with experts and users about the problem-solving aspects of a KBS.

According to CommonKADS, a CME is a three layer model: the lower level is called the "Domain Layer" since it contains all the concepts and the logical relations between them, the middle level is called the "Inference layer" because it contains inference steps using the domain rules, attributes the semantic roles the domain concept must play in an inference step and organizes inferences and roles in inference structures, and the upper layer, the "Task Layer" defines the methods (i.e. the prototypical algorithms) that can be used to achieved some problem solving goals with the use of the corresponding inference structure. As a result, a CommonKADS Conceptual Model of Expertise describes the cognitive process an expert uses to solve a given problem with the corresponding domain knowledge. One fundamental property of a CommonKADS CME is that the internal structures of a MCE is independent of the expertise domain because it is directly linked with a cognitive task (Diagnosis for example). So CommonKADS provide a set of expertise templates defining a domain schema, an inference structure and the corresponding method for the main cognitive tasks. Such templates being generic (i.e. independent of the domain knowledge), they can be used to accelerate the acquisition and the modeling of the experts' knowledge given a cognitive task.

The TOM4D methodology aiming at producing the generic model of a dynamic process an expert uses to produce the domain knowledge in order to solve a problem, TOM4D completes the CommonKADS methodology: a TOM4D model can be used to validate the domain knowledge of a CommonKADS Conceptual Model of Expertise or to help its construction. Inversely, it is clear that the use of the domain knowledge of a CommonKADS CME will accelerate the construction of a TOM4D model of the corresponding process. Unfortunately, it is rare that such a CommonKADS Conceptual Model of Expertise is available when modeling a dynamic process. But the CommonKADS templates can always be used to facilitate the interpretation of the available knowledge about a dynamic process. The "Knowledge Interpretation" step of the TOM4D methodology uses precisely the CommonKADS templates to this aim: this constitutes then an important tool to control the introduction of the semantics in the TOM4D modeling process.

But these templates being domain independent, they are not sufficient to provide a correct physical interpretation to a set of variables. To this aim, the TOM4D methodology resorts to the Tetrahedron of States (ToS) [28, 32] formalized by Rosenberg and Karnopp [29] in the early 80s (Fig. 7.7). The ToS is a framework that describes a set of four generalized continuous variable ("e" for "effort", "f" for "flow", "p" for "impulse" and "q" for "displacement") linked together with a set of five binary relations (Fig. 7.7). These binary relations are represented with equations defining three type of constant ("C" for "Capacity", "R" for "Resistance" and "I" for "Inductance"), and one differential operator. The three binary relation, namely "q = C · e", "e = R · f" and "p = I · f" defines the algebraic relation between the five generalized variables and the two others defines theirs temporal behavior. In other words, these five binary relations define three types of generic components.

The major interest of the ToS is that this framework is common to any physical domains (electromagnetism, fluid dynamics, thermodynamics, etc.) of the classical newtonian physics. For example, in the hydraulic dam domain, the "Hydraulic ToS" maps the "water flow" to the generalized flow "f", the "volume" to the to generalized displacement "q", the "pressure" to the generalized effort "e" and the hydraulic momentum to the generalized impulse "p". This way, any variable x_i of the set X of the perception model $PM(X(t))$ will receive a physical dimension and as a consequence, any proposition about the relations between two hydraulic variables can be analyzed according to the "Hydraulic ToS": a proposition that don't satisfy one of the five relations of the Hydraulic ToS must be rejected by the knowledge

Fig. 7.7 Tetrahedron of States (ToS) (based on [28, p. 1728])

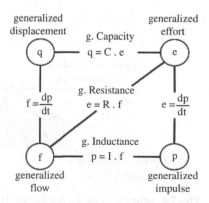

engineer. The ToS constitutes then a powerful tool of the TOM4D methodology to interpret and represent the expert knowledge, that is to say to control the introduction of the semantics within the TOM4D modeling process.

In complementary with the ToS, the Formal Logic framework, and more precisely the first order predicate calculus in particular, is also used by the TOM4D methodology as a resource to provide a way to analyze the logical consequences of a ToS validated proposition: a new proposition can satisfy the semantic constrains of the ToS but can be contradictory with the current version of the model. In this latter case, the knowledge engineer must solve the contradiction either with the rejection of the new proposition or with an adequate update of the current version of the model that contains the new proposition. It is to note that using the Formal Logic framework allows the utilization of Reiter's Theory of Diagnosis [11] to facilitate the analysis of the model coherence.

Finally, the combination of the CommonKADS templates, the ToS and the first order predicate calculus constitute a powerful framework to control the introduction of semantic elements during the modeling process of the TOM4D methodology.

The knowledge representation formalisms of the TOM4D methodology are presented in the next chapter throughout the illustration of the application of the methodology to a real-world example of dynamic process, the French Sapins dam. The interested reader is invited to refer to [3] to get compact introduction of the TOM4D knowledge representation formalisms.

7.6 Application to the French Sapins Dam

The French Sapins dam is a homogeneous dam with a granite arena structure on a granite foundation, 16 m high, and impounding a 2 hm^3 lake [7].

It has a vertical drain and a horizontal drain at the interface of the foundation with the downstream half of the dam (Fig. 7.8). The top of the drain was lower than the normal reservoir level. The upstream shoulder is instrumented with three pore pressure cells. Vertical drain discharge is recorded.

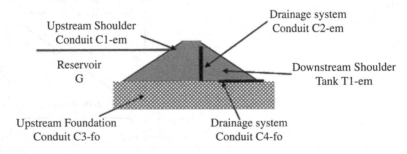

Fig. 7.8 Structure of the French Sapins' Dam

Reservoir filling began in November 1978. The piezometry records show that the line of saturation in the upstream fill rose higher than the vertical drain as early as July 1980, indicating that water seeped into the downstream shoulder.

Drain outflow initially decreased in November 1980. Following this, partial emptying of the reservoir occurred in November 1981, the drain outflow increased in December 1981 and remained stable until October 1982. In October 1982, the drain outflow began to decrease again until to July 1883. In September 1988, a very wet area was noted at the downstream toe. In mid-October 1988, the damp path grew larger and localized slides and muddy seepages could be observed on the upstream shoulder. For safety reasons, it was decided to completely empty the reservoir and improve its structural safety before bringing the dam back into service. Figure 7.9 represents a part of the degradation scenario of the French Sapins' dam (more details can can be found in [7]).

The investigations about the Sapin's dam concluded that a mechanism of internal erosion was operative. The particle size grading of the fill material made it particularly sensitive to internal erosion, and this led to the gradual clogging of the vertical drain, aided by the fact that the drain did not meet standard filter rules. Clogging caused first the upstream fill to become gradually saturated and then the downstream fill as the infiltration water overtopped the drain. Saturation of the downstream fill material diminished its engineering properties, leading to shallow slides. The flowing water started to carry away the fines to the point where piping would have quickly developed and caused complete dam failure if drawdown had not been ordered quickly.

7.6.1 Knowledge Interpretation

The first step of the TOM4D methodology modeling process, the "Knowledge Interpretation" step, produces the scenario model $M(\Omega) = <SM(\Omega), FM(\Omega), BM(\Omega)>$ from the degradation scenario of Fig. 7.9.

The interpretation of the knowledge has been made with the utilization of a conceptual template inspired from Sachem's conceptual template [13]. Figure 7.10 provides the inference structure allowing the recognition of process phenomena from the timed measures provides by sensors. In this figure, the static knowledge is hidden in order to make it clearer. The concepts of this cognitive template allow to analyze the available knowledge, mainly a chapter of Peyras' PHD memory (cf. [7]). Sensors provides the timed measures that the "fit" inference validates and orders to make signals (cf. Fig. 7.9). Event signals are geometrical patterns characterizing some specific behavior breakdown that can be detected on the curves made from signals with signal processing algorithms. Signal phenomenon are temporal pattern of particular event signal describing a specific evolution of the signal over time. Such an evolution results of the matching of the flow of event signals against behavioral models. Process phenomenon result of the aggregation over time and space of signal phenomenon to describe an abstract behavior of the

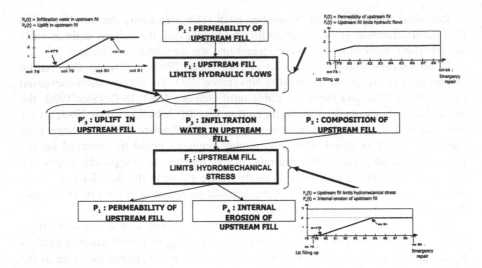

Fig. 7.9 Sapin's Dam degradation scenario (from [7])

Fig. 7.10 Sachem inspired
inference structure (from [13])

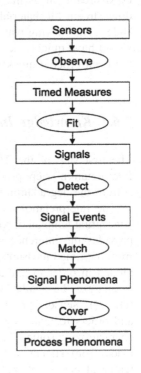

process so that the observed signal phenomenon are explained by a process phenomenon (i.e. a current set of process phenomenon "cover" the current set of signal phenomenon).

7.6.2 Perception Model

The construction of $M(\Omega)$ is detailed in [4]. This model has lead to define the global process of Sapin's dam as the process $P(t)$ that relates three time functions $V(t)$, $Q_s(t)$ and $Q_f(t)$. These functions are linked together with the hydraulic ToS of Fig. 7.11a: $V(t)$ denotes a volume (in m^3), Q_s and Q_f being respectively the output and the leaking water flow in the drainage system. The hydraulic ToS allows also to show that the experts neglects the momentum of pressure $Pp(t)$ to analyze the risk of a dam breakdown. This leads to the adapted ToS according to the model $M(\Omega)$ (Fig. 7.11b).

One fundamental contribution of the "knowledge Interpretation" step of the TOM4D methodology is the interpretation of the notion of dam breakdown: the drainage system resistance is represented with the time function $R(t)$ meaning that the leaking water flow can evolve from 0 (no leaking, $R(t) \rightarrow \infty$, the drainage system is filled) to a maximum value corresponding to a dam breakdown (i.e. $R(t) = 0$). The dam breakdown risk can then be formally represented with the following equation:

$$\forall t, \exists t_k, t \geq t_k, R(t) = 0 \tag{7.4}$$

This equation means that there is a time value t_k that stamps the dam breakdown (i.e. $R(t_k) = 0$): this situation is clearly the one to avoid. It is to note that, according to scenario model $M(\Omega)$, the resistivity of the civil structure is not an observable function of the dam. The Eq. 7.4 is then an indirect way to assess the dam breakdown risk. It show also the fundamental importance of the diagnosis task in the risk assessment.

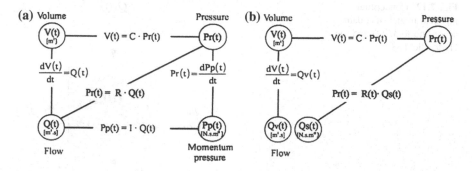

Fig. 7.11 **a** Hydraulic ToS **b** and its adaptation to $M(\Omega)$

The adapted hydraulic ToS (Fig. 7.2b) also leads to define the experts conceptual perception of the dam (Fig. 7.12) under the form of a water column of a constant capacity C (cf. Fig. 7.11), closed by a porous cork and fed with an input water flow $Qv(t)$. The cork play the conceptual role of a resistance $R(t)$ the value of which evolves over time. The output water flow $Qs(t)$ is then linked with the volume of water column by the ToS equations of Fig. 7.11b.

In this conceptual perception of a dam, the column of water behaves like a perfect cylinder: not leaking and no deformation phenomenon (i.e. an indeformable and waterproof tube). This means that:

$$\forall t, Qf(t) = 0 \qquad (7.5)$$

In practice, the permeability of the dam structure degrades with its aging and some times, according to the faults during the construction, a sufficient permeability is not assured. This leads to water leaking phenomenon ($Qf(t) > 0$) that degrades the dam structures and can provoke its destruction (i.e. $R(t) \rightarrow 0$).

The simple model defines clearly the abstraction level dam experts use to manage their diagnosis reasoning: the process is made of a set of three time functions $X(t) = \{V(t), Qs(t), Qf(t)\}$.

Formally, the TOM4D methodology defines the perception $PM(X(t))$ is a 3-tuple $<X, \Psi, R^p>$ where (cf. [3] for a complete presentation of the knowledge representation formalism):

- X is a set of variable names. Here $X = \{V, Qs, Qf\}$.
- Ψ is a set of constant values defining ranges for the variables of X. Typically, $\Psi = \{\psi_i\}$ is a set of thresholds ψ_i allowing the application of the discretization principle of the Timed Observations Theory. In the example, Ψ is made of height thresholds (Table 7.1) defining three ranges for V and Qs (low, normal, high levels) and two for Qf (zero, not-zero).
- $R^p = R^g * R^n * R^a$ is a set of predicates that relate the elements of X with the elements of Ψ in order to determine constrains on variables. R^g is a set that describes the operating goals. R^n and R^a are sets that specify the normal and

Fig. 7.12 Conceptual physical model of a dam according to $M(\Omega)$ and the Hydraulic ToS

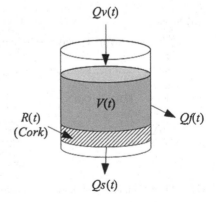

Table 7.1 Thresholds Ψ^{x_i}, Constants δ^{x_i} and Observation Classes C^{x_i} for each x in $X = \{V, Qs, Qf\}$

Variable	Ranges	Constants	Obs. Classes
V	$V < \psi^{f_1}$	δ^{v_0}	$C^{v_1} = \{(V, \delta^{v_0})\}$
	$\psi^{v_1} \leq V < \psi^{v_2}$	δ^{v_1}	$C^{v_2} = \{(V, \delta^{v_1})\}$
	$V \geq \psi^{v_2}$	δ^{v_2}	$C^{v_3} = \{(V, \delta^{v_3}\}$
Qs	$Qs < \psi^{s_1}$	δ^{s_0}	$C^{s_1} = \{(Qs, \delta^{s_0})\}$
	$\psi^{s_1} \leq Qs < \psi^{s_2}$	δ^{s_1}	$C^{s_2} = \{(Qs, \delta^{s_1})\}$
	$Qs \geq \psi^{s_2}$	δ^{s_2}	$C^{s_3} = \{(Qs, \delta^{s_3})\}$
Qf	$Qf < \psi^{f_1}$	δ^{f_0}	$C^{f_1} = \{(Qf, \delta^{f_1})\}$
	$Qf \geq \psi^{f_1}$	δ^{f_1}	$C^{s_2} = \{(Qf, \delta^{f_1})\}$

abnormal operating modes, respectively. In the example, the exploitation goal is defined as "no leaking", the normal operating mode define either the empty dam or the normal working dam. The abnormal operating mode is defined as any other mode but those defined in R^n.

- $R^g = \{\forall t \in R^+, Qf(t) < \psi^f{}_1\}$ //no leaking
- $R^n = \{\exists t_0 \in R, \forall t \in R^+, t < t_0, V < \psi^f{}_1 \wedge Qs < \psi^s{}_1 \wedge Qf < \psi^f{}_1$ //Empty dam
  $\exists t_0 \in R, \forall t \in R^+, t \geq t_0, \psi^v{}_1 \leq V < \psi^v{}_2 \wedge \psi^s{}_1 \leq Qs < \psi^s{}_2 \wedge Qf < \psi^f{}_1$ //Normal dam
  $\}$

Figure 7.12 and Table 7.1 resume the perception model $PM(X(t))$ of the hydraulic dam. It is to note that the input volume $Qv(t)$ is not contained in $X(t)$: this means that the risk assessment about a dam is not concerned with the way the dam is filled.

The perception model considers a dam as a structure having one input, corresponding to the water flow Q_v of Fig. 7.12 (which is not under consideration in the risk assessment of the dam destruction), one internal variable V and two outputs that are linked with two water flow Qs and Qf (Fig. 7.13).

The functioning of this structure is provided by the adapted hydraulic ToS of Fig. 7.11b under the form of two equations:

$$\forall t, \ dV(t) / dt = Qv(t) - (Qs(t) + Qf(t)) \tag{7.6}$$

Fig. 7.13 Conceptual structure of a dam according to the perception model

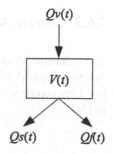

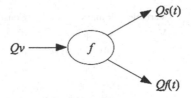

Fig. 7.14 Conceptual functioning of a dam according to the perception model

$$\forall t, V(t) = C \cdot R(t) \cdot Qv(t) \tag{7.7}$$

The relation between the values of $V(t)$ and those of $Q_s(t)$ and $Qf(t)$ are then provided by the following differential equation:

$$\forall t, Qs(t) + Qf(t) = Qv(t) - C \cdot d(R(t) \cdot Qv(t))/dt \tag{7.8}$$

The capacity C of the water column being, by construction, a constant, this equation establishes clearly the relation between the values of the output flows Qs and Qf with the input flow Qv, according to the evolution of the cork permeability R (t) (Fig. 7.14). The cork permeability being an internal variable, $R(t)$ is not observable: assessing the value of $R(t)$ is then clearly the object of the risk analysis of the dam breakdown. This shows clearly the importance of the diagnosis task to the risk assessment.

According to the control theory of dynamic system, $R(t)$ is an internal variable of the dam component of Fig. 7.12 that play the rôle of state variable. This means that the relations between the values of V, Qs and Qf depends strongly on the evolution of $R(t)$. In other words, there is not a unique behavioral model that can render the behavior of the Eq. 7.8.

A deeper analysis is then required to precise the relation between the evolution of $R(t)$ and the model of the dam the experts uses to assess the dam destruction risk of Sapins' dam (Fig. 7.8). This analysis constitutes the last step of the TOM4D modeling process, the "Generic Modeling". The next section describes this last step and will show the utilization of the perception model during the whole modeling step.

7.6.3 Generic Modeling

The "Generic Modeling" step aims at building the structural model $SM(X(t))$, the functional model $FM(X(t))$ and the behavioral model $BM(X(t))$ corresponding to the previous perception model $PM(X(t))$ (details of the modeling process can be found in [5]).

Figure 7.8 shows that the structure of the Sapins' dam is made of six components: a vertical drainage system (c_1), a downstream shoulder (c_2), an upstream shoulder (c_3), a reservoir (c_4), an upstream foundation (c_5) and a horizontal drainage system (c_6). Some variables can be directly assessed using data collected on dams. For instance, the volume $V(t)$ for the upstream shoulder is assessed using the water level. These data are of two types [33, 34]: visual observations (presence of seepage, presence of sinkhole, etc.), monitoring data (piezometry, outflow measurement, etc.). It is to note that the input flows Qv of each components are not assessed.

The key point of this analysis has been to consider each of these six components as a special instance of the generic component describes by the Figs. 7.14 and 7.15 and the Eq. 7.8 and to describe the dam with a set of binary relations between them. This lead to a new definition of the set X of variable that contains now $6*3 = 18$ variables V_i, Qs_i and Qf_i, i = 1...6.

The TOM4D structural model $SM(X(t))$ aims at describing the structural relations between these six components. $SM(X(t))$ is a 3-tuple $<COMPS, R^i, R^x>$ where (Fig. 7.15):

- COMPS = $\{c_1, c_2, c_3, c_4, c_5, c_6\}$ is a set of conceptual components c_i corresponding to the "water column" of Fig. 7.12. A conceptual component is a set input and outputs:

  - $c_i = \{in(c_i), out_1(c_i), out_2(c_i)\}$ //c_i is made of one input (Qv) and two outputs (out_1 for Qs and out_2 for Qf).

- $R^i = \{out_1(c_4) = in(c_5), out_1(c_4) = in(c_3), out_1(c_5) = in(c6), out_1(c_5) = in(c_6), out_2(c_6) = in(c_2), out_1(c_3) = in(c_1), out_2(c_3) = in(c_2), out_2(c_1) = in(c_2), out_1(c_2) = in(c_6)\}$, is a set of binary "equal" predicates of the form $out_k(c_i) = in(c_j)$ where c_i and $c_j \in$ COMPS, $k \in \{1, 2\}$, linking one of the two outputs of a component c_i to the input of a component c_j.

- $R^x = \{out_1(c_5) = Qs_5, out_2(c_5) = Qf_5, out_2(c_4) = Qs_5, out_2(c_6) = Qf_6, out_1(c_3) = Qs_3, out_2(c_3) = Qf_3, out_2(c_1) = Qf_1, out_1(c_2) = Qs_2\}$ is a set of binary "equal" predicates of the form $out_k(c_i) = x_j$ where $c_j \in$ COMPS, $k \in \{1, 2\}$, and $x_j \in X$. This set associates a variable x_i to one and only one components c_i of C (Qv_i is associated with the output of c_i, Qs_i is associated with the first input of c_i, and Qf_i with its second input).

Figure 7.15 proposes a graphical representation of $SM(X(t))$. This structural model can be interpreted as the experts' conceptual structure of the Sappins' dam: this model defines the interactions between the six component. The formal representation can be used to model the dam according to Reiter's logical theory of Diagnosis.

The functioning and the behavior of the dam is then logically constrained by the relations of this structural model $SM(X(t))$. If the six components have the same (generic) structure, their functioning and their behavior differs according to their role in the dam structure. The analysis of the available knowledge identifies 4 types of components:

- "Conduit": Horizontal and vertical drainage systems (c_6 and c_1), upstream shoulder (c_3) and foundation (c_5) belongs to this type. The two first are

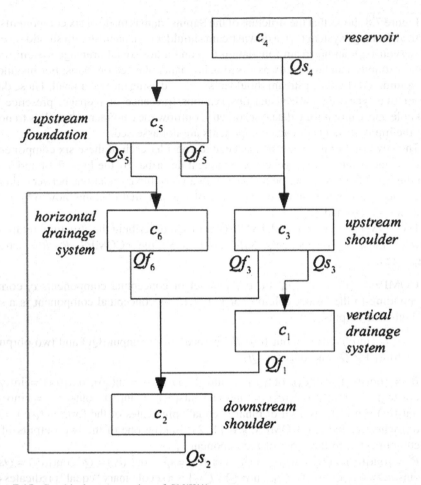

Fig. 7.15 Graphical representation of $SM(X(t))$

characterized by $R(t) \rightarrow 0$ in normal operations (i.e. such conduits aims at collecting all the input flow Qv, cf. Eq. 7.9 which typically represents a pipe), the two others being characterized by $R(t) \rightarrow R_0.R_0$ is a constant corresponding to a low value: the permeability of such conduits tends to be constant in order to control the output flow Qs according to the value of the product of the capacity C with the permeability R_0 (cf. Eq. 7.10 which typically represents a low pass filter).

$$\forall t, Qs(t) + Qf(t) = Qv(t) \tag{7.9}$$

$$\forall t, Qs(t) + Qf(t) = Qv(t) - C \cdot R_0 dQv(t)/dt \tag{7.10}$$

- "Tank": the downstream shoulder (c_2) belong to this type of component. Such a component is characterized by $R(t) \to \infty$ in normal operations, meaning that its output flow $Qs(t) \to 0$ (the output flow of a tank cannot be negative). This means that the only output flow is $Qf(t)$ (cf. Eq. 7.11).

$$\forall t, Qs(t) = Qf(t) \tag{7.11}$$

- "Generator": The reservoir (c_4) belongs to this type that is characterized by $R(t) = 0$ (no permeability), and $\forall t$, $Qf(t) = 0$ and $Qs(t) \neq 0$ (i.e. the internal volume is never null whatever are $Qv(t)$ and $Qs(t)$. This means that the role of a generator is to maintain a minimal internal volume $V(t)$ so that the output flow is never equal to zero.

$$\forall t, Qs(t) > 0 \tag{7.12}$$

The four Eqs. 7.9–7.12 defines the functioning of the different components a dam is made with. The functioning of the dam results both of these equations and the way the components are connected together. The TOM4D methodology represent the functioning of a structure with a functional model $FM(X(t))$, which is a 3-tuple $<\Delta, F, R^f>$ where:

- $\Delta = *\Delta_{xi}, \forall x_i \in X$ (cf. column "Constant" of Table 7.1).
- $F = \{f_i : \Delta^n \to \Delta / \forall x \in X^n, f_i(x) \in \Delta\}$ //F is a set of scalar functions defined on Δ^n.
- $R^f = \{x_i = f_i(x_1, x_2, \ldots, x_n)\}$ is a set of equality predicates defining the value of a variable x_i of X according to the values of a subset of $X - \{x_i\}$.

According to the spatial discretization principle (Fig. 7.4) used to represent the knowledge, logical relations can be defined for the values of the variables. Generally speaking a set $\Delta_{xi} = \{0, 1, 2\}$ of three discrete values are defined for the variables x_i of X (details about the spatial discretization applied to model Sapins' dam with TOM4D can be found in [33, 35] in particular). According to the knowledge and the four Eqs. 7.9–7.12, a table of values is associated with the functions of the six components of $SM(X)$. This leads to the functional model of Fig. 7.16.

It is to note that the functional model of each of the four types of components (i.e. "Conduit", "Generator" and "Tank") can be directly used as generic models in a first order logic theory. In other words, the dam structural model of Fig. 7.15 and the four generic functional models of the "Conduit", the "Generator" and the "Tank" can be directly translated in a "model" according to Reiter's logical theory of Diagnosis. In other words, a model according to Reiter's logical theory of Diagnosis contains only the structural and the functional models according to the TOM4D methodology.

The Eq. 7.10 illustrates the dynamic property of the upstream foundation and the upstream shoulder. The tables of value defined for these two components correspond to the application of a discretization principle to steady states or equilibrium

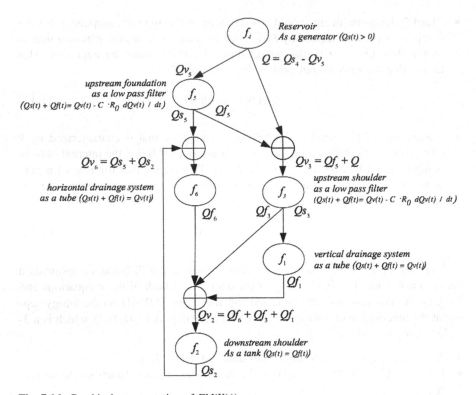

Fig. 7.16 Graphical representation of $FM(X(t))$

of these components. This justifies the need for a complementary model that will be able to describe this dynamics throughout the evolutions of the values of the variables.

This description is the role of the behavioral model. A TOM4D behavioral model is defined on the notion of "discernible state": a temporal binary relation $r_{ij}(C_i, C_j, \left[\tau_{ij}^-, \tau_{ij}^+\right])$ between two observation classes $C_i\alpha\{(x_i, \delta_i)\}$ and $C_j\alpha\{(x_j, \delta_j)\}$ defines a unique "discernible state" s_{ij}. Such a state is called "discernible" because it is characterized uniquely with a discernible state vector X defined on the cartesian product $\Pi\Delta_{xi} = \Delta_{x1} \times \Delta_{x2} \times \cdots \times \Delta_{xn}$, $\Delta_{xi} = \{\delta_k^i\}$, $k = 1\ldots m$, of the definition domain $\Delta_{xi} = \{\delta_k^i\}$ of all the variables x_i of $X = \{x_1, x_2, \ldots, x_n\}$.

A TOM4D behavioral model $BM(X(t))$ is a 3-tuple $<S, C, R^c>$ where:

- $S = \{s_{ij}\}$ is a set of discernible states s_{ij} defined on $C \times C$. An identification function denoted $i(s_{ij})$ defined on S associates a unique value X_{ij} defined on X so that, $\forall s_{ij}, s_{nm} \in S, s_{ij} = s_{nm} \Leftrightarrow i(s_{ij}) = i(s_{nm})$ (i.e. $X_{ij} = X_{nm}$).

    - $i: S \rightarrow X$, $i(s_{ij}) = X_{ij}$, $X_{ij} \in X$.

- $C = \{C_i\}$ is a set of observation classes $C_i = \{(x_i, \delta_i)\}$, $\delta_i \in \Delta_{xi}$, defined as singletons.
- $R^c = \{s_{jl} = \gamma(C_l, s_{ij})\}$ is a set of equality predicates linking a state s_{ij} to its successor s_{jl} under the condition of the observation of an occurrence $C_l(t_k)$ of the class $C_l \in C$.

Such a behavioral model $BM(X(t))$ aims at being implemented on a classical Finite State Machine (FSM) that utilizes the relations $\gamma(C_l, s_{ij})$ to determine the value s_{ij} of the current discernible state $s(t)$ according to the timed observations $o(k)\alpha(\delta_i, t_k)$ that occurs over time. Knowing $s(t)$, the identification function $i(s(t))$ provides the corresponding value of the vector $X(t)$ so that a unique value is assigned to each variables x_i of X.

Given a sequence $\omega = \{o(k)\}$ of timed observations $o(k)\alpha(\delta_i, t_k)$, a transition from the state s_{ij} to a state s_{jl} is fired *iff*:

- the current state $s(t)$ of the Finite State Machine implementing $BM(X(t))$ is in the state s_{ij}: $s(t) = s_{ij}$;
- the current timed observation $o(k)\alpha(\delta_l, t_k)$ of ω belongs to the class $C_1 = \{x_i, \delta_l\}$;
- R^c contains one predicate $s_{jl} = \gamma(C_l, s_{ij})$.

When these three conditions are satisfied, the new value of the current state $s(t)$ is provided with the transition function σ defined on $S \times \omega \times R^c \rightarrow S$ of the Finite State Machine such as, at time $t = t_k$, $s(t) = \sigma(s_{ij}, (\delta_l, t_k), \gamma(C_l, s_{ij}))$.

A behavioral model can be build for the four types of components (the two types of "Conduits", the "Tank" and the "Generator", cf. Figs. 7.17, 7.18 and 7.19 respectively). For each of them, the discernible state vector X is define on the generic set of variables $X = \{V, Qs, Qf\}$. According to the discretization principles utilized by the experts, the corresponding set of C of observation classes is made of singletons. For example, the following behavioral models have been made with a set $\Delta_{xi} = \{0, 1, 2\}$ of three discrete values are defined for the variables x_i of $X = \{V, Qs, Qf\}$.

In the Figs. 7.17 ($BM(Tank)$), 7.18 ($BM(Generator)$) and 7.19 ($BM(Conduit)$), the classes are denoted "C_{xy}" where x denotes the index of a variable in the vector X (i.e. $x = 1$ for V, $x = 2$ for Qs and $x = 3$ for Qf) and y denotes the index of a value (i.e. a constant) in $\Delta_{xi} = \{0, 1, 2\}$ (i.e. $y = 0$ for $\delta = 0$, 1 for $\delta = 1$ and 2 for $\delta = 2$). The states are denoted "s_j" where the index j is the value associated with the corresponding discernible state. To each state s_j is associated the value of the vector X defined on the generic set of variables $X = \{V, Qs, Qf\}$: for example, the value "$\{0, 0, 1\}$" associated with s_0 means: $V = 0$, $Qs = 0$ and $Qf = 1$.

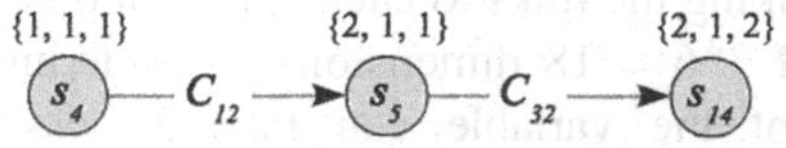

$$\{1, 1, 1\} \qquad \{2, 1, 1\} \qquad \{2, 1, 2\}$$

$$\boxed{s_4} - C_{12} \rightarrow \boxed{s_5} - C_{32} \rightarrow \boxed{s_{14}}$$

Fig. 7.17 Graphical representation of the Generic BM(Tank)

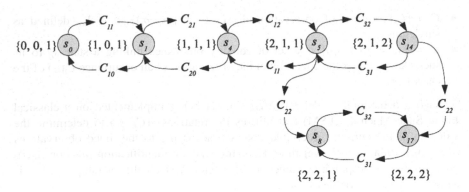

Fig. 7.18 Graphical representation of the Generic *BM*(Generator)

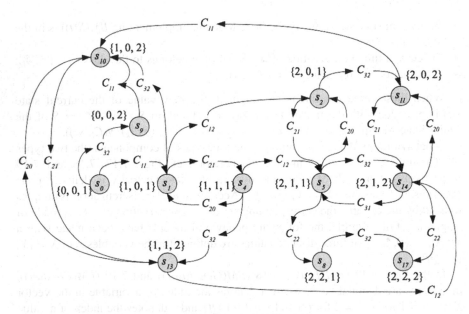

Fig. 7.19 Graphical representation of the Generic *BM*(Conduit)

The global behavior model of a dam is a combination of the behavioral model of one generator, one tank and four conduits. It is then obvious that this model is very complicated because it contains not only the seven states of BM(Generator) (Fig. 7.18), the 4*12 = 48 states of the four BM(Conduit) (Fig. 7.19) and the three states of BM(Tank) (Fig. 7.17), that is to say 7 + 48 + 3 = 58 discernible states, but also all the elements making the links to each states. In other words, the discernible state vector is made of 3*6 = 18 dimensions: considering the set of 3 constant $\Delta_{xi} = \{0, 1, 2\}$, each of the variable can take 3 possible values leading to $3^{18} = 3,874,204,489$ states in the dam discernible state space! It is then clear that the building of the "*BM(Dam)*" behavioral model is not realizable at hand.

This explains why we developed an algorithmic approach to assess the risk. This approach is based on a new diagnosis algorithm the presentation of which is out of the scope of this chapter. The interested reader is invited to read the specialized papers [36, 37] for a detailed presentation).

7.7 Diagnosis-Based Risk Assessment with TOM4D Models

The algorithm proposed in [33, 36, 37] applies the decomposition principle proposed in Sect. 7.3.2. It is based on a two level diagnosis reasoning: one for each component c_i of COMPS and the second for the dam. The proposed algorithm implements a decentralized and incremental approach of diagnosis [38–41] for quantized dynamic system [42].

The diagnosis reasoning for a component (the generator, the tank of one of the four conduit for the Sapins' dam example) uses the corresponding behavioral model to determine the current state of the concerned component according to the available timed observations. Next, using the corresponding perception model, the current state of each component is classified as normal or abnormal and the consequences of the determination of the current state for each component are computed at the dam level with the utilization of the functional model (Fig. 7.16 for the for the Sapins' dam example). As a consequence, the assessment of the risk breakdown of a dam results of the diagnosis of each of its components. Details of the execution of the diagnosis algorithm can be found in [33]. This section aims at illustrating, by the mean of Sapins' dam example, the way the risk of a dam breakdown can be assessed when using the diagnosis algorithm and the TOM4D models described in the preceding sections.

The degradation scenario of the Sapins' dam (cf. the introduction of Sect. 7.6) is made the following events and their representation according to the functional model:

1. November 1978, reservoir c_3 is filled: V_3 (November 1978) = 2;
2. July 1980, saturation in the upstream fill rose higher than the vertical drain c_1: Qs_1 (July 1980) = 0;
3. December 1981, the drain outflow increased: Qs_1 (December 1981) = 1;
4. July 1983, the drain outflow decreased: Qs_1 (July 1983) = 0;
5. September 1988, detection of a very wet area in the downstream shoulder c_2: Qf_2 (September 1988) = 2.

The observation of this latter event has entailed the decision of emptying the Sapins' dam because of the too high level of destruction risk. This suite of events composes a sequence of timed observations $\omega = \{ C^3{}_{12}$ (2, November 1978), $C^1{}_{20}$ (0, July 1980), $C^1{}_{21}$ (1, December 1981), $C^1{}_{20}$ (0, July 1983), $C^2{}_{32}$ (2, September 1988)}. In this sequence, an observation class is denoted $C^i{}_{xy}$ where i denote the index of the component c_i in the structural model of Fig. 7.15, x and y being respectively the index of a variable in the vector X (i.e. x = 1 for V, x = 2 for Qs and

$x = 3$ for Qf) and the index of a value (i.e. a constant) in $\Delta_{xi} = \{0, 1, 2\}$ (i.e. $y = 0$ for $\delta = 0$, 1 for $\delta = 1$ and 2 for $\delta = 2$)). The decompositional principle allows so decompose the superposition $\omega \cup \omega_1 \ast \cup \omega_2 \cup \omega_3$ in three sequences according to the component a timed observation is concerned with:

1. $\omega_1 = \{C^1{}_{20} \ (0, \text{July } 1980), C^1{}_{12} \ (1, \text{December } 1981), C^1{}_{20} \ (0, \text{July } 1983)\}$;
2. $\omega_2 = \{C^2{}_{32} \ (2, \text{September } 1988)\}$;
3. $\omega_3 = \{C^3{}_{12} \ (2, \text{November } 1978)\}$.

The first timed observation of ω_1, $C^1{}_{20}$ (0, July 1980), concerns the output flow Qs of the vertical drainage system c_1 that behaves like a "Conduit": according to the behavioral model of a conduit (Fig. 7.19), this timed observation indicates that the state of the vertical drainage system can be: $s^1{}_1$, $s^1{}_2$, $s^1{}_{10}$ or $s^1{}_{11}$. The next timed observations of ω_1, $C^1{}_{21}(1, \text{December } 1981)$ and $C^1{}_{20}$ (0, July 1983) don't change this evaluation.

The unique timed observation of ω_2, $C^2{}_{32}$ (2, September 1988), concerns the leaking flow of the upstream shoulder c_2 that behaves like a "Conduit": the state of the upstream shoulder c_2 can be: $s^2{}_9, s^2{}_{10}, s^2{}_{11}, s^2{}_{13}, s^2{}_{14}$ or $s^2{}_{17}$.

The unique timed observation of ω_3, $C^3{}_{12}$ (2, November 1978), concerns the downstream shoulder c_3 that behaves like a "Tank": according to the behavioral model of a generic tank of Fig. 7.17, the current discernible state of the reservoir can only be $s^3{}_5$.

The discernible state identification function $i(s)$ provides, for all state of all the components, the values of the vector X corresponding to each states. For example, $i(s^3{}_5) = (2, 1, 1)$ meaning that $V^3 = 2$, $Qs^3 = 1$ and $Qf^3 = 1$. Doing so for each state, the logical relations of the functional model allows to reduce the global diagnosis to the following final diagnosis: $s^1{}_1, s^2{}_{14}$ and $s^3{}_5$. This diagnosis corresponding to the following situation: $V^1 = 1$, $Qs^1 = 0$ and $Qf^3 = 1$ (the vertical drainage system is filled), $V^2 = 2$, $Qs^2 = 1$ and $Qf^3 = 2$ (the downstream shoulder is leaking) and $V^3 = 2$, $Qs^3 = 1$ and $Qf^3 = 1$ (the volume in the upstream shoulder is high). According to the experts, this situation correspond to two potential phenomena: the internal erosion through the embankment and its sliding of the downstream shoulder. The combination of such phenomena constitutes a major incident, which could have led to the failure of Sapins' dam. Sapins' dam has been rehabilitated in 1989.

The situation of September 1988 was due to the particle size grading of the fill material, particularly sensitive to internal erosion. It led to the gradual clogging of the vertical drainage system, which did not comply standard filter rules.

7.8 Conclusion

The risk assessment relies fundamentally on a diagnosis task.

Such a highly cognitive task requires a model of the process to be diagnosed. The main problem with such an approach is the complexity of the model of the process and its adequacy with the diagnosis outputs. This difficulty increases when diagnosing a dynamic process, notably because the knowledge of the behavior of a dynamic process is difficult to constitutes (long time learning) and to explicit (the behavioral knowledge is mainly based on sub-symbolic piece of knowledge that are mostly impossible to formulate with words).

This chapter presents a Knowledge Engineering methodology called TOM4D (Timed Observations Modeling for Diagnosis) that is based on the Timed Observation Theory [1]. This theory constitutes a mathematical framework for modeling dynamic processes from timed data by combining the Markov Chain Theory, the Poisson Process Theory, the Shannon's Communication Theory and the Logical Theory of Diagnosis. It provides then technical basis of the TOM4D methodology. In particular, the Timed Observation Theory provide the concepts of variable, timed observation and observation classes that are particularly suited to model dynamic processes.

A TOM4D model of a dynamic process is made of four models: a structural model containing the knowledge about the components and their connection, a functional model containing the knowledge about the values the process variable is made with, the behavioral model containing the knowledge about the state transitions that governs the evolution of the values of the process variable over time, and finally, a perception model that defines the operating goal of the dynamic process and its normal and abnormal operating modes. The coherence of the relations between these models is provided by the Timed Observation Theory's notion of variable.

The TOM4D methodology is a syntax-driven approach of Knowledge Engineering. This means that the introduction of the semantic elements during the modeling is controlled with three types of interpretation models: CommonKads templates, formal logic and the tetrahedron of state. This means that the purpose of the TOM4D methodology is to complete the conceptual model of expertise one can made with a Knowledge Engineering methodology like CommonKADS. For example, the domain knowledge contained in a CommonKADS conceptual model of expertise constitutes logical consequences of the corresponding TOM4D model of the dynamic process.

The aim of this chapter is to show that the TOM4D methodology constitutes a great help to acquire the knowledge about the assessment of the risks with the operation of a dynamic process. TOM4D is illustrated with a real world dynamic process, the French Sapins' hydraulic dam the destruction of which has been avoided in 1988. The chapter shows that the TOM4D models capture the knowledge at the same level of abstraction than the experts so that a diagnostic algorithm has been defined to facilitate the utilization of the TOM4D models. One of the main advantage of this approach is to avoid the building of huge behavior1a model for a dam: only the knowledge of the behavior of its components is required. The state space of the French Sapins' dam being very huge, this example constitutes a good illustration of the operational flavor of the TOM4D methodology and the proposed diagnosis algorithm to the risk assessment about the operation of a dynamic process.

The future developments of the presented works will be concerned with the investigation of more complex properties concerning diagnosis-based risk assessment on one hand, notably with the introduction of the abstraction method proposed in [43, 44], and the development of automatic modeling tools based on the Timed Observation Theory. The aim of these latter is to facilitate the syntactic and the semantic coherence of a model under construction.

References

1. Le Goc, M. (2006). *Notion d'observation pour le diagnostic des processus dynamiques: Application à Sachem et à la découverte de connaissances temporelles*. Habilitation à Diriger des Recherches. Unversité de Droit d'Economie et des Sciences d'Aix-Marseille.
2. Pomponio, L., & Le Goc, M. (2010). Timed observations modelling for diagnosis methodology: A case study. In J. A. M. Cordeiro, M. Virvou, & B. Shishkov (Eds.), *ICSoft 2010—Proceedings of the 5th International Conference on Software and Data Technologies* (pp. 504–507). Athens: SciTePress.
3. Le Goc, M., Masse, E., & Curt, C. (2008). Modeling processes from timed observations. In *Proceedings of the 3rd International Conference on Software and Data Technologies (ICSoft'08)* (pp. 249–256).
4. Le Goc, M., & Masse, E. (2007). Towards a multimodeling approach of dynamic systems for diagnosis. In *Proceedings of the 2nd International Conference on Software and Data Technologies (ICSoft'07)* (pp. 277–282).
5. Pomponio, L., & Le Goc, M. (2013). Integrating knowledge engineering with knowledge discovery in database: TOM4D and TOM4L. In C. Faucher & L. Jain (Eds.), *Recent advances in knowledge engineering: Paradigms and applications*. New York: Springer.
6. CIGB. (1994). *Bulletin 93—ageing of dams and appurtenant works—review and recommendations*. Paris: CIGB.
7. Peyras, L., Royet, P., & Boissier, D. (2006). *Dam ageing diagnosis and risk analysis: Development of methods to support expert judgment*. NRC Canada.
8. Foster, M., Fell, R., & Spannagle, M. (2000). A method for assessing the relative likelihood of failure of embankment dams by piping. *Canadian Geotechnical Journal, 37*, 1025–1061.
9. Farinha, F., Portela, E., Domingues, C., & Sousa, L. (2005). Knowledge based systems in civil engineering: Three cases studies. *Advances in Engineering Software, 36*, 729–739.
10. Shannon, C. E. (1948). A mathematical theory of communication. *The Bell System Technical Journal, 27*(379–423), 623–656.
11. Reiter, R. (1987). A theory of diagnosis from first principles. *Artificial Intelligence, 32*, 57–95.
12. Dagues, P. (2001). Théorie logique du diagnostic à base de modèles. *Diagnostic, Intelligence Artificielle, et Reconnaissance des Formes* (pp. 17–105). Hermes Science Publications.
13. Le Goc, M. (2004). SACHEM, a real-time intelligent diagnosis system based on the discrete event paradigm. *Simulation, 80*, 591–617.
14. Le Goc, M., & Ahdab, A. (2012). *Learning Bayesian networks from timed observations*. Saarbrücken: LAP LAMBERT Academic Publishing GmbH & Co. KG.
15. Benayadi, N., & Le Goc, M. (2010). Mining timed sequences with TOM4L framework. In *Proceedings of the 12th International Conference on Enterprise Information Systems (ICEIS 2010)*(pp. 111–120).
16. Ahdab, A., & Le Goc, M. (2010). Learning dynamic Bayesian networks with the TOM4L process. In *Proceedings of the Fifth International Conference on Software and Data Technologies (ICSoft 2010)*(pp. 353–363).

17. Bouché, P., Le Goc, M., & Coinu, J. (2008). A global model of sequences of discrete event class occurrences. In *Proceedings of the Tenth International Conference on Enterprise Information Systems (ICEIS 2008)*(pp. 173–180).
18. Nonaka, I. (1994). Dynamic theory of organizational knowledge creation. *Organization Science, 5*, 14–37.
19. Nonaka, I. (1991). The knowledge-creating company. *Harvard Business Review, 69*, 96–104.
20. Alavi, M., & Leidner, D. E. (2001). Review: Knowledge management and knowledge management systems: Conceptual foundations and research issues. *MIS Quarterly, 25*, 107–136.
21. Polanyi, M. (1966). *The tacit dimension*. New York: Doubleday & Company, Inc.
22. Nonaka, I., & Konno, N. (1998). The concept of "Ba": Building a foundation for knowledge creation. *California Management Review, 40*, 40–54.
23. Studer, R., Benjamins, V. R., & Fensel, D. (1998). Knowledge engineering: Principles and methods. *Data and Knowledge Engineering, 25*, 161–197.
24. Damasio, A. (2005). *Descartes' error: Emotion, reason, and the human brain*. Putnam, 1994, revised Penguin edition, 2005.
25. Damasio, A. (1999). *The feeling of what happens: Body and emotion in the making of consciousness*. New York: Harcourt.
26. Schreiber, G., Akkermans, H., Anjewierden, A., et al. (2000). *Knowledge engineering and management: The CommonKADS methodology*. Cambridge: MIT Press.
27. Breuker, J., & de Velde, W. V. (1994). *CommonKADS library for expertise modelling*. Amsterdam: IOS Press.
28. Chittaro, L., Guida, G., Tasso, C., & Toppano, E. (1993). Functional and teleological knowledge in the multimodeling approach for reasoning about physical systems: A case study in diagnosis. *IEEE Transactions on Systems, Man and Cybernetics, 23*, 1718–1751.
29. Rosenberg, R. C., & Karnopp, D. C. (1983). *Introduction to physical system dynamics*. New York: McGraw-Hill.
30. Chittaro, L., & Ranon, R. (1999). Diagnosis of multiple faults with flow-based functional models: The functional diagnosis with efforts and flows approach. *Reliability Engineering and System Safety, 64*, 137–150.
31. Zanni, C., Le Goc, M., & Frydman, C. (2006). A conceptual framework for the analysis, classification and choice of knowledge-based diagnosis systems. *KES-International Journal of Knowledge-Based & Intelligent Engineering Systems, 10*, 113–138.
32. Chittaro, L., & Ranon, R. (1996). Augmenting the diagnostic power of flow-based approaches to functional reasoning. In *AAAI-96 Proceedings* (pp. 1010–1015).
33. Fakhfakh, I., Curt, C., Le Goc, M., & Torres, L. (2012). Diagnosis of the hydraulic dam safety based on multi-modeling approach. In *18eme Congrès de Maîtrise des Risques et Sûreté de Fonctionnement*.
34. Curt, C., Peyras, L., & Boissier, D. (2010). A knowledge formalization and aggregation-based method for the assessment of dam performance. *Computer-aided Civil and Infrastructure Engineering, 25*, 171.
35. Fakhfakh, I., Curt, C., Le Goc, M., & Torres, L. (2012). Using the multimodeling approach to model dam behavior. *Journal of Computing in Civil Engineering*.
36. Fakhfakh, I., Le Goc, M., Torres, L., & Curt, C. (2012). Modeling and diagnosis dynamic system from timed observations. In *The 23rd International Workshop on the Principles of Diagnosis (DX'2012)*.
37. Fakhfakh, I., Le Goc, M., Torres, L., & Curt, C. (2012). Toward a decompositional and incremental approach of diagnosis dynamic system from timed observations. In *The 23rd International Workshop on the Principles of Diagnosis (DX'2012)*.
38. Benveniste, A., Haar, S., Fabre, E., & Jard, C. (2005). Distributed monitoring of concurrent and asynchronous systems. *Discrete Event Dynamic Systems, 15*(1), 33–84.
39. Cordier, M.-O., & Grastien, A. (2007). Exploiting independence in a decentralised and incremental approach of diagnosis. In *20th International Joint Conference on Artificial Intelligence*.

40. Grastien, A., Cordier, M., & Largouöt, C. (2005). Incremental diagnosis of discrete-event systems. In *DX'05, International Workshop on Principles of Diagnosis* (pp. 119–124). Pacific Grove, California, USA.

41. Pencolé, Y., Cordier, M.-O., & Rozé, L. (2001). Incremental decentralized diagnosis approach for the supervision of a telecommunication network. In *Twelfth International Workshop on Principles of Diagnosis (DX-01)* (pp. 151–158).

42. Lunze, J. (1999). Discrete-event modelling and diagnosis of quantized dynamical systems. In *DX-99, 10th International Workshop on Principles of Diagnosis* (pp. 147–154). Loch Awe, Ecosse, Royaume Uni.

43. Pomponio, L., Le Goc, M., Anfosso, A., & Pascual, E. (2012). Levels of abstraction for behavior modeling in the GerHome project. *International Journal of E-Health and Medical Communications, 3,* 12–28.

44. Pomponio, L., Le Goc, M., Pascual, E., & Anfosso, A. (2011). Resident's activity at different abstraction levels: Proposition of a general theoretical framework. In *The 6th IEEE International Conference on Intelligent Data Acquisition and Advanced Computing Systems: Technology and Applications, IDAACS'2011* (pp. 540–545). Prague, Czech Republic.

Chapter 8
Building a Domain Ontology to Design a Decision Support Software to Plan Fight Actions Against Marine Pollutions

Jean-Marc Mercantini

Abstract The purpose of this chapter is to show that cognitive approaches can offer very powerful engineering environments to tackle issues raised by risk management. The investigated issue is the planning of actions to fight accidental marine pollutions. The response proposed is a software tool (GENEPI) to support stakeholders to plan fight actions during emergency situations or crisis management with the objective to minimize pollution impacts. From a methodological perspective, the chapter shows the importance to develop ontologies (i) for structuring a domain as perceived by its actors and (ii) for building computer tools aimed to support problem solving in that domain. Such tools are imbued of the knowledge shared by the actors of the domain, what make them more effective within critical situations. The methodological process is based on the Knowledge Engineering method: "Knowledge Oriented Design" (KOD), which is founded on the fields of linguistics and cognitive anthropology. The resulting ontology, the architecture of the software tool and the plan generation mechanism are presented and discussed.

8.1 Introduction

That the issue of risk is addressed from a social, technological or anthropological point of view, it turns out that its perception, understanding, acceptance and reactions that may arise are strongly related to socio-cultural background of an individual or a group of individuals. Moreover, the evolution of the activities of our modern societies, due to the globalization of markets and the use of new information technologies, is responsible for situations more complex involving the cooperation and collaboration of individuals from different cultures (e.g. social

J.-M. Mercantini (✉)
Laboratoire des Sciences de l'Information et des Systèmes,
Domaine Universitaire de Saint Jérôme, Avenue Escadrille Normandie-Niemen,
13397 Marseille Cedex 20, France
e-mail: jean-marc.mercantini@Lsis.org

© Springer-Verlag Berlin Heidelberg 2015   197
J.-M. Mercantini and C. Faucher (eds.), *Risk and Cognition*,
Intelligent Systems Reference Library 80, DOI 10.1007/978-3-662-45704-7_8

background, professional background, academic background) and the use of "intelligent" machines. If this complexity can be controlled in "normal" situations, it can become a real source of danger in critical situations where decisions must be taken and executed under high stress. In this context, the design of new software tools to support solving safety issues must take into account the experience and vision of all actors according to the issues raised by the complexity of critical situations.

One of the main problems that arise in the design of new computer tools to support the resolution of safety problems is linked to a non-stabilized vocabulary, reflecting semantic and conceptual differences within the community concerned by the tool. These differences can be found at a cultural level within the actors involved in a specific situation as well as within existing methodological tools or, within the different disciplinary sectors. The notion of ontology and the works currently developed by the community of cognitive scientists and knowledge engineers can provide relevant answers to the raised problems.

The term *ontology* is often associated to the knowledge related to objects of a delimited universe and their relations. Ontology refers to a conceptual language used for the description of this delimited universe (domain). A domain ontology is an example of knowledge level model [1]. The emergence of this notion in Knowledge Base System (KBS) engineering comes from the fact that the way to observe the world and its interpretation are directly dependent of the observer culture, his (her) means to observe it as well as to his (her) intentions. One of the objectives of ontologies is to facilitate the exchange of knowledge between humans, between humans and machines as well as humans via machines [2]. In this sense, it becomes necessary to resolve the difficulties caused by observation, representation and interpretation of (normal or critic) situations to facilitate problem solving (intent).

Works developed in the context of this chapter refers to the elaboration of ontologies in order (i) to provide answers to problems caused by work situations involving the cooperation and collaboration of individuals from different cultures (e.g. social background, professional background, academic background) as well as (ii) to implement intelligent machines in the domain of safety and risk management. Indeed, if these problems can be controlled during "normal" workload situations, they can become real sources of danger during critical situations where decisions must be taken and implemented under tight constraints.

Considering the scenario where ontology must serve as base in the specification of a computer tool aimed to assist users to solve a safety problem within a given domain, we argue that the ontology should not be limited to structure that domain, but it should also structure it taking the problem and the solving method used into account. In this sense, it is important that the method used for building ontologies helps the knowledge engineer in conceptualizing the Td triplet: <Domain, Problem, Method>. Ontologies should ensure the coherence of this triplet in order to serve as base in designing computer tools. From our point of view, an important role of ontology is on one hand to express the objects of the universe on which the reasoning will be made (domain representation) but also, on the other hand, to express the objects used in describing the problem and those used by the solving method. The objective of this approach is to design methodological and software tools imbued of the knowledge shared by the domain stakeholders, what make them more effective, especially within critical or crisis situations.

The engineering method that implements this approach is presented in the chapter. It is based on the elaboration of a domain ontology in order to design a problem solving software to assist stakeholders to plan fight actions against marine pollutions in the Mediterranean area. This problem solving software, called "Generation of Intervention Plans Module" (in French "GENEration de Plans d'Intervention": GENEPI), is a specific module from a wider research program called CLARA 2 ("Calculs Liés Aux Rejets Accidentels": Calculations Relating to Accidental Releases in the Mediterranean).

The following research fields are concerned by this chapter: (1) from the maritime field perspective, the chapter presents a software tool to assist crisis management staff to minimise pollution impact and (2) from a methodological perspective, the chapter shows the importance of developing ontologies (i) for structuring a domain (at the conceptual level) as perceived by its actors and (ii) for using these ontologies in building computer tools dedicated to assist human actors in solving complex problems in that domain. The Knowledge Oriented Design method (KOD) [3], originally designed to develop KBSs, has been used to elaborate the domain ontology.

After describing the CLARA 2 project and the functioning principles of the GENEPI module (Sect. 8.2), a general overview of the approaches commonly used for the construction of ontologies is given (Sect. 8.3) and the general criteria that characterize a method for ontologies development will be exposed. The presentation of the KOD method will focus on the main elements that satisfy these general criteria (Sect. 8.4). It will be presented, then, the ontology building process, which integrates the three steps of the KOD method (Sect. 8.5). To validate our approach, results obtained from the study of the process of generating intervention plans will be presented (Sect. 8.6). Finally, we conclude on the suitability of this method for building ontologies and the influence of the corpus on the process modelling (Sect. 8.7).

8.2 The CLARA 2 Project

Although the Mediterranean is only one hundredth of the sea surface it supports 30 % of the volume of international maritime traffic. An estimated 50 % of goods transported could present a risk to different degrees. A study on shipping accidents in the Mediterranean sea [4], conducted between 1977 and 2003 identified 376 accidents involving hydrocarbons and 94 accidents involving hazardous and noxious substances (HNS). These accidents have resulted in a total discharge of 305,000 tons of hydrocarbons and 136,000 tones of HNS. These events highlight the criticality of the risk induced by transport activities in that maritime area.

In general, the strategy of fight against marine pollution from hydrocarbon following shipping accident is divided into three complementary stages: (i) stopping the pollution source, (ii) recovery of the maximum volume of hydrocarbon on the sea and, when the pollutant reached the coast, (iii) cleaning the polluted coastline.

The fight strategies against chemical products depend on their physico-chemical properties and the hazardous effects caused by the accident (fire, explosion and/or dispersion of toxic clouds). There are many intervention techniques to combat pollution and their effectiveness depends on the situations in which they are implemented. Thus, it appears that the choice of a fight technique in a response plan is not trivial and requires taking into account a large number of parameters.

In a general way, decisions and actions undertaken by operational center actors need to mobilize a large number of information from various sources and under high time pressure. These information need to be integrated in a coherent way prior to be interpreted and finally to be the base of any decision and action. Among the main activities carried out by operational center actors it can be cited: situation analysis, determining fight strategies, choosing the right fight actions, elaborating fight action plans and anticipating future situations.

Stakeholders usually implied in an operational center for managing maritime accidents are: the Navy, the National Administrations, the local administrations, the National Meteorology and expert institutes like the French Research Institute for Exploitation of the Sea (IFREMER) or the Centre of Documentation, Research and Experimentation on Accidental Water Pollution (CEDRE). Managing such accidents generates complex and critical work situations. The complexity is characterized by: shared knowledge, shared competences, cultural distances between actors or stakeholders, no coherent vocabulary and no coherent computer systems. The criticality is mainly characterized by: time constraints, climate constraints and environment vulnerability (natural and human).

The project CLARA 2 (Calculations Relating to Accidental Releases in the Mediterranean) brings responses to these problems. It aims to develop and implement a computer tool to assist operational center stakeholders to manage situations resulting from a maritime accident having caused a spill of pollutants, whether of chemical or petroleum. To carry out this national project (funded by the Research National Agency), a consortium of 13 partners was formed [5].

CLARA 2 (Fig. 8.1) should facilitate the rapid establishment of relevant exclusion zones to alert, but also to protect people, goods and environment, to mobilize appropriated fighting means and to anticipate critical situations. It also provides information on the capabilities of bio-accumulate in the food chain substances released and a preliminary approach to risk in terms of toxicological effects on humans is proposed in case of atmospheric dispersion of toxic gases. The software tool is based on a simulator designed to predict the location of a pollutant, changes in its concentration in the sea and in the atmosphere following a massive spill. It helps to know the effects in the case of a fire, provides information on the bioaccumulation capacity of some marine organisms and provides sensitivity indicators according to the polluted areas. A Geographical Information System (G.I.S) provides coastal and

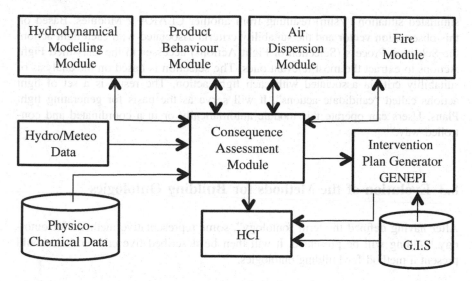

Fig. 8.1 Architecture of the CLARA 2 software tool

vulnerability maps. In addition, CLARA 2 generates plans on the steps to take and methods of intervention to implement (the generation module of intervention plans: GENEPI).

The fighting plans generated by GENEPI take account of the accidental situations and their changes over time. The set of methods and intervention techniques that could be mobilized have been classified and suitability criteria with situations have been established and associated to each of them. Figure 8.2 shows the functioning principle of this module. Access to the GENEPI module is done through an observation vector of a real situation (Vreal) and/or an observation vector of a

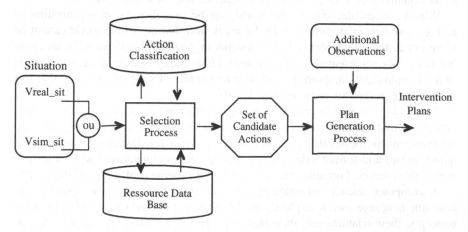

Fig. 8.2 Data flow diagram of the GENEPI module

simulated situation (Vsim) resulting from another CLARA 2 Modules. Based on this observation vector and the suitability criteria associated with each fight action, the Selection Process (Selection of Fight Actions) accesses to the Classified Fight Actions to extract the most relevant ones. The selection is based on the analysis of suitability criteria associated with each fight action. The result is a set of fight actions called "candidate actions". It will serve as the basis for generating fight Plans. Users can operate this module automatically or in a coordinated and controlled way.

8.3 Evolution of the Methods for Building Ontologies

After having defined the term "ontology", some representative methods for ontology building will be presented. It will then be described five criteria that should present a method for building ontologies.

8.3.1 Definitions

Ontology is the branch of philosophy known as "the science of being" [6, 7] that studies the nature of being or kinds of existence, and seeks to unify abstractions into an overall philosophy of what it means to exist [7]. First defined in Philosophy as a fundamental branch of Metaphysics, Ontology studies being or existence and their basic categories and relationships, to determine what entities and what types of entities exist [8]. In formalizing the nature of things and the distinctions between them, Ontology is applied to fields such as Theology, Information Science and Artificial Intelligence [7]. Briefly, it means that the Ontology field studies the world as an organization of its fundamental categories and their interrelations.

Within a computer system, the world can be modelled as an organization of entities and their relationships, with the restriction that the whole world cannot be represented, but only a part of it. For systems in Artificial Intelligence "what exists is what can be represented", in this context, an ontology is an explicit specification of a conceptualization, defined as an abstract simplified view of the world that one wishes to represent for some purpose [9, 10].

An ontology is also "a program defining a set of representational terms", where definitions associate the names of entities in the universe of discourse (e.g. classes, relations, functions, or other objects) with human-readable text describing what the names mean, and formal axioms that constrain the interpretation and well-formed use of these terms. Formally, an ontology is the statement of a logical theory [9].

In computer science, an ontology can thus be defined as a conceptual representation language used to explicitly model the world studied (or part of it) by using concepts, their relations and their meanings. In other words, an ontology can be defined as an operational tool for knowledge representation, as also proposed by Bachimont [11]: "ontologies provide notional and conceptual resources to formulate knowledge and make it explicit".

An ontology can also be defined by its form. Indeed, it can be presented as a set of terms naming the entities composing the universe of discourse (classes, relations, functions, etc.) [9]. This aspect was also studied by Uschold and Grüninger [2] who stated that an ontology may take several forms, but it always includes a vocabulary of terms and a specification of their meaning, this specification including definitions, relations between concepts, and links imposing a structure on the domain to constrain possible interpretations of terms. Uschold and Grüninger [2] also pointed out that the vocabulary and the meaning can be expressed at different degrees of formality: highly informal (expressed loosely in natural language), semi-informal (expressed in a restricted and structured form of natural language), semi-formal (expressed in an artificial formally defined language), rigorously formal (meticulously defined terms with formal semantics, theorems and proofs of such properties as soundness and completeness).

Ontology can also be defined in terms of the roles it plays. One role is to support knowledge sharing activities [9]. It means that ontology can provide a common language to people that need to collaborate, for instance by exchanging and understanding information. This role was described by Bachimont [11] as a shared working framework, that different actors can mobilize and in which they can find an anchoring point. Another role of ontology is to represent the meaning of what is exchanged between information systems [11]. These two roles were also included in the definition given by Uschold and Grüninger [2]: an ontology is a unifying framework for different viewpoints and can serve as the basis for enabling communication (between people, between people and systems, and between systems), and this unifying conceptual framework is intended to function as a lingua-franca. For Maedche [12], ontologies are models used to communicate meaning.

8.3.2 Ontology Engineering

Ontology Engineering studies the process of building ontologies, which is presently an important research field. Ontology Engineering is a multidisciplinary field borrowing several concepts from other disciplines, for instance Knowledge Engineering, Artificial Intelligence, Knowledge Management, Computational Linguistics, Information Systems, Information Retrieval and Extraction, Information Integration, Database; Ontology Engineering can be defined as a branch of Knowledge Engineering that exploits principles of formal Ontology to build ontologies [13, 14].

At the beginning of the nineties, ontologies were recognized as an important issue for enabling knowledge sharing [9, 15, 16]. Since then, several methodologies have been proposed for building ontologies, whatever from scratch, by reuse, manual, automatic, etc. Methods can be ascendant or bottom-up (by abstracting data), descendant or top-down (by specializing concepts) or middle-out (by specializing and abstracting most important concepts into other concepts). Those methods have been studied in several overviews for instance Corcho et al. [17], Fernández-López and Gómez-Pérez [18], Fernández-López [15] and Pinto and Martins [16]. Among representative methodologies, one can mention TOVE [19], ENTERPRISE [20, 21], METHONTOLOGY [15, 22] and TERMINAE [23].

TOVE was one of the first methods to be proposed and it showed how one can formalize ontologies directly from the knowledge of the domain studied. With ENTERPRISE, it was experimented how it was possible to start with the expert knowledge in order to create a conceptual model of the ontology by generalization and specialization, before the formal and implementation models. The purpose of the METHONTOLOGY method was to express ontologies at the knowledge level and give tools to help one build them. With METHONTOLOGY, the process of ontology building was situated in a framework combining the principles of Knowledge Engineering and Software Engineering. METHONTOLOGY allows one to build ontologies thanks to an intermediary conceptual model, without the necessity of an a priori expert knowledge. The ontology construction is achieved during a life cycle consisting in four steps: specification (identification of the purpose and users of the aimed ontology), conceptualization (structuring the domain knowledge at the knowledge-level), formalization (automated translation of the conceptual model into a formal model) and implementation (writing the formal model with an implementation language). Guidelines were proposed in the form of predefined tables in order to facilitate acquisition and conceptualization of knowledge. A software tool, called ODE [24], was developed to guide the use of METHONTOLOGY. Finally, with TERMINAE, the ontology building process was situated in a framework of Knowledge Engineering exploiting linguistics principles. The TERMINAE methodology is composed of four main steps: corpus constitution, linguistic study, normalization and formalization. During corpus constitution, the relevant documents are selected. The linguistic step allows identification of the terminology from the corpus thanks to the use of natural language processing (NLP) tools. During the normalization step, the expert of the domain chooses the concepts in order to structure and validate the knowledge that will be represented in the ontology. This step is achieved based on the results produced by NLP tools and the expert knowledge. The normalization step corresponds to the conceptualization step in METHONTOLOGY. TERMINAE is also a software tool to guide the expert in the ontology building process.

State of the art on ontologies and on Ontological Engineering shows that many research projects have been achieved and many others are actually conducted in order to develop new computer tools and methods to help build ontologies. To the best of our knowledge, there are no completely automated tools that can, from a corpus describing a domain, build an ontology of this domain. Present techniques can be qualified as semi-automated and require the intervention of an expert in Ontology Engineering for relevant results. Most of the linguistic approaches proposed the integration of several existing tools, which raises the problem of coherence when results are produced by the enchainment of different tools.

The most critical steps are knowledge acquisition from documentary sources and domain structuring with the identification of semantic relations.

Research work in Ontological Engineering also put in evidence the necessity to elaborate guidelines for ontology building at the knowledge-level from documentary sources, which can be multimodal. The first ontologies were built from scratch and without any methodology or software engineering standard like the development life cycle to guide their development; after some initial experiments on ontology building, a few design criteria were found [9]. METHONTOLOGY and TERMINAE provide the ontology builder with a method, a tool and some explicit guidelines. However, much work stills needs to be done in order to develop more explicit guidelines.

Research work in Ontology Engineering has put in evidence 5 main steps for building ontologies [25]:

1. Ontology Specification. The purpose of this step is to provide a clear description of the problem studied as well as the method to solve it. This step allows one to describe the objectives, scope and granularity size of the anticipated ontology.
2. Corpus Definition. The purpose is to select among the available information sources those that will allow the objectives of the study to be attained. This step has been described for instance in the TERMINAE methodology [23].
3. Linguistic Study of the Corpus. This step consists in a terminological analysis of the corpus in order to extract the candidate terms and their relations. This step has been proposed for instance with TERMINAE [23] and the method of Dahlgren [26]. Linguistics is specially concerned by the ontology concept to the extent that available data for ontology building are often expressed as linguistic expressions [23]. The characterization of the sense of these linguistic expressions leads to determine contextual meanings.
4. Conceptualization. Within this step, the candidate terms and their relations resulting from the linguistic study are analysed. The candidate terms are transformed into concepts and their lexical relations are transformed in semantic relations. The result of this step is a conceptual model. This step was proposed with the following methodologies: ENTERPRISE [2, 21], METHONTOLOGY [15, 22], TERMINAE [23].
5. Formalization. The purpose of this step is to express the conceptual model with a formal language. This step was proposed with the following methodologies: ENTERPRISE [20, 21], METHONTOLOGY [15, 22], TERMINAE [23].

8.3.3   Criteria Needed for a Methodology for Building Ontologies

The literature about ontologies reported in the preceding paragraphs, highlights a number of properties characterizing ontologies. These properties must be taken into account by the methods intended to assist in their construction. The main capabilities required by these methods are: (i) linguistic capabilities, (ii) conceptualization capabilities, (iii) capabilities to model reasoning and capabilities to express (iv) the generic features of concepts and (v) the consensual features of concepts.

The linguistic capability is due to the fact that ontology is defined primarily as a language. Therefore, a method for developing such a language must necessarily make available linguistic tools to help ontologists build this language.

The conceptualization capability comes from the fact that this language is qualified as conceptual, in the sense that no term in this language is related to a specific object but to a concept clearly defined. Therefore a method for developing such a language must provide aid tools for the conceptualization.

The ability to capture the reasoning comes from the formal character of ontologies, allowing them to support reasoning operations that could be mobilised during solving problem. These reasoning operations can be organized into problem solving methods. Therefore, an elaboration method of ontologies has to offer tools for the capture, the modelling and the formalization of these reasoning operations.

The capability to express the generic nature of the concepts comes from the fact that ontologies are distinct conceptual models [27], among others, by the generic nature of the concepts which constitute an ontology. Indeed, the scope of an ontology is not confined to a particular application, but on the contrary, it must cover a wide scope. Therefore, it is important to dispose a method capable of yielding the generic nature (and not specific) of the conceptual language to build.

An ontology is by definition consensual, it is important that an elaboration method of ontologies could encourage, thanks to its approach, the emergence of this consensual characteristic.

8.4 KOD: A Knowledge Engineering Method

8.4.1 General Presentation of KOD

KOD belongs to the family of KE methods designed to guide the knowledge engineer in the task of KBS elaboration. This method was designed to introduce an explicit model between the formulation of the problem in natural language and its representation in the formal metalanguage chosen. KOD is based on an inductive approach, which requires to explicitly express the cognitive model (also called the conceptual model) based on a corpus of documents, comments and experts speeches.

The main features of this method are based on linguistics and cognitive principles. Its linguistics bases make it well suited for the acquisition of knowledge expressed in natural language. Thus, it proposes a methodological framework to guide the collection of terms and to organize them based on a terminological

Table 8.1 KOD, the three modelling levels

Paradigms models	Representation	Action	Interpretation
Practical	Object static representation: taxeme	Dynamic representation of active objects: acteme	Inferences
Cognitive	Object static organization according to theirs properties: taxonomy	Dynamic object organization: actinomy	Reasoning pattern
Software	Classes	Methods	Rules

analysis (linguistic capacity). Through its anthropological bases, the knowledge engineer has a methodological framework facilitating the semantic analysis of the terminology used to produce a cognitive model (conceptualization capacity). It guides the work of the knowledge engineer from the extraction of knowledge up to the development of the conceptual model.

The implementation of the KOD method is based on the development of three successive models: the practical models, the cognitive model and the software model (Table 8.1). Each of these models is developed according to the three paradigms: <Representation, Action, Interpretation/Intention>.

The Representation paradigm gives the KOD method the ability to model the universe such as experts represent it. This universe is made of concrete or abstract objects in relation. The KOD method provides methodological tools to develop the structure of this universe of knowledge according to this paradigm. The Action paradigm gives the KOD method the ability to model the behaviour of active objects that activate procedures upon receipt of messages. Thus, the action plans designed by human operator, as well as those of artificial operators, will be modelled in the same format.

The Interpretation/Intention paradigm gives the KOD method the capability to model the reasoning used by experts to interpret situations and elaborate action plans related to their intentions (reasoning capacity).

The practical model is the representation of a speech or document expressed in the terms of the domain, by means of taxemes (static representation of objects), actemes (dynamic representation of objects) and inferences (base of the cognitive structure of the task). A taxeme is a minimum grammatical feature; it is the verbalisation of an object or a class of objects. An acteme is the verbalisation of an act or a transformation, a unit of behaviour. An inference is the act or process of deriving logical conclusions from premises known or assumed to be true. The cognitive model is constructed by abstracting the practical models. The cognitive model is composed of taxonomies, actinomies and reasoning patterns. The software model results from the formalization of a cognitive model expressed in a formal language independently of any programming language.

8.4.2 KOD: Elaboration of the Models

The first step in modelling is the development of practical models, which consist, based on a corpus Mp (Fig. 8.3), in extracting the terms and the relations linking them through a terminological analysis to provide a terminological language. The terms of the language are classified into <taxemes, actemes, inferences> consistent with the three paradigms of the method (Table 8.1). The taxemes make it possible to represent objects and concepts regarded by experts of the domain. The actemes make it possible to describe activities carried out by experts of the domain and causing changes in state-level objects. The inferences are the atomic elements by which experts are building their reasoning to interpret a situation and to plan their action. At the end of the first step of the modelling process, each document of the corpus $(D_1, D_2, ..., D_n)$ is modelled by means of a Practical Model $(MP_1, MP_2, ..., MP_n)$ (Fig. 8.3). Each Practical Model is a representation of a specific case of the problem.

The second step consists in developing the cognitive model related to the problem, from the set of practical models. The process consists in: (i) analyzing

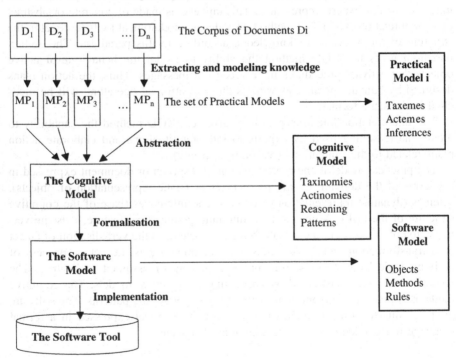

Fig. 8.3 The KOD design process of software systems to assist managers in problem-solving and decision-making

synonyms and homonyms terms occurring in the practical models; (ii) transforming the resulting terms into concepts and determining the right identifier; and (iii) transforming the lexical relationships into semantic ones. In accordance with the three paradigms of the method, the cognitive model is organized as follows <taxonomies, actinomies, reasoning patterns> (Fig. 8.3). The taxonomies result from the taxeme classification. It is presented as a hierarchical tree structure showing the connections between each concept and related object. The actinomies are the result of an orderly organization of actemes defining an action plan. The reasoning patterns are defining a structure for modelling the expert reasoning when he plans his actions previously to execute them. Thus, the process implemented encourages the emergence of generic and consensual characteristics of the conceptual language developed during this second stage. Indeed, solving problems of synonyms and homonyms promotes consensual characteristics and abstractions of practical models promote obtaining a conceptual and generic language.

The third stage concerns the development of the software model that requires to previously express the conceptual language by means of a formal language. The choice of the formal language depends on the properties of the conceptual model and it must be adapted to the nature and complexity of concepts to be represented. The formalization operation consists in integrating the elements of the conceptual model in the definition of classes and objects to constitute a formal model that will be used for the software development. The implementation phase that follows consists in translating the formal model by means of a programming language.

8.4.3 The Life Cycle of Ontologies with KOD

The development of ontologies can be seen as a knowledge-intensive task. In this sense, the use of KOD seems to be a priori pertinent because one of the aims of a KE methodology is to help define knowledge of a domain studied. In this section, we explain how KOD can be integrated in the process of developing ontologies (Table 8.2).

The projection of the KOD method on the general approach for developing an ontology shows that KOD guides the corpus constitution and provides the tools to meet the operational steps 3 (linguistic study) and 4 (conceptualization). In what follows, each step of the life cycle of developing an ontology is put in correspondence with the process proposed by KOD. More specifically, it will be developed: (i) the corpus building; (ii) the linguistic study; and (iii) the conceptualization.

Under previous researches, the KOD method has been already implemented [28–30] in the domains of road safety, safety of urban industrial sites and study of conduct errors of industrial plants.

Table 8.2 Integration of the KOD method into the elaboration process of ontologies

Elaboration process of ontology	KOD process	Elaboration process of ontology with KOD
1. Specification		1. Specification
2. Corpus definition		2. Corpus definition
3. Linguistic study	1. Practical model	3. Practical model
4. Conceptualisation	2. Cognitive model	4. Cognitive model
5. Formalisation		5. Formalisation
	3. Software model	6. Software model

8.5 Elaboration of the Ontology for GENEPI

8.5.1 Ontology Specification

The KOD method does not offer tools facilitating the specification of ontology. To carry out this step, many authors recommend the use of the concept of scenario [2, 25, 31] with the objectives to clarify and justify the validity of building ontology, the future uses and the future addressees. We do not further develop this stage but we illustrate it by giving summaries of the scenario that have been drafted within the framework of the triplet Td: <Domain, Problem, Method>.

The domain is that of maritime accidents with the release of pollutant products (hydrocarbon or chemical) and causing a marine pollution. The problem is to assist crisis management teams to elaborate action plan to fight the pollution. The problem solving method consists in the elaboration of a cooperative software tool, which implement the generation process of fight actions against marine pollutions. The method can be divided into three tasks: (i) specifying the human-computer cooperation process, (ii) specifying the generation process of the set of candidate fight actions and (iii) specifying the generation process of a plan.

8.5.2 Corpus Definition and Analysis

Definition and analysis of the corpus are performed on the basis of the specification of the ontology as well as the consideration of the properties of practical and conceptual models resulting from the application of the KOD method. Thus, the documents to be collected must be both representative of the triplet <Domain, Problem, Method> and meet the criteria of suitability required by the three paradigms <Representation, Action, Interpretation/Intention>. The combination of the triplet (Td) with the three paradigms constitutes a helpful grid to analyse the ontology specification with the goal to define the documents that must constitute the corpus.

The corpus is based on documents from CEDRE (Centre of Documentation, Research and Experimentation on Accidental Water Pollution), from REMPEC

(REgional Marine Pollution Emergency REsponse Centre for Mediterranean sea), from CEPPOL (Centre of Practical Expertise in Pollution Response), and from the maritime prefecture. The subjects of these documents are related to maritime accidents, to the implementation of fight actions, to provisional organisation as well as to the census of human and material means. The Table 8.3 presents a general overview of the corpus in relation with the searched knowledge.

The ORSEC plans provide a global and static representation of the out of context fight organization. They define the framework within which the GÉNÉPI module can be activated. They also provided us with the available and mobilizable resources and their distribution on the Mediterranean coast.

The synthesis documents relating to the descriptions of fight actions provide very precise information about their out of context implementation. Thus, we could identify the means necessary for the implementation of these actions, the adequacy criteria (out of context) and applicability rules.

Table 8.3 Kind of documents that constitute the corpus of the study

Kind of documents	Document content	Kind of searched knowledge
ORSEC plans (French emergency plans)	Global and static representation of the fight organization (out-of-context)	Representation of the out-of-context knowledge universe (Taxonomies)
		Out-of-context resources (Taxonomies)
		Planned fight strategies (reasoning framework)
The return on operating experience documents about accidents (CEDRE):	Global description of the in context fight organization (resources, actions, strategies)	Representation of the in-context knowledge universe (taxonomies)
The Prestige accident		Accident scenarii (Taxonomies, Actinomies, reasoning frameworks)
The Erika accident		Generic accident scenarii (Taxonomies, Actinomies, reasoning frameworks)
The ECE sinking		Resources (Taxonomies)
The wreck of DOLLY		In-context suitability criteria (Taxemes)
...		Fight strategies (reasoning frameworks)
Method and technic descriptions: the REMPEC guides, the fight data sheets from CEDRE and the chemical intervention guidelines from CEDRE	Specific and detailed analysis of the out of context fight actions	Implementing protocol of fight actions (Actinomies)
		Specific resources (Taxinomies)
		Intrinsic suitability criteria (Taxemes)
		Applicability rules (Inferences)
Quality procedures (from CEDRE) for crisis or accidental event management	Organization of crisis management, task description and distribution between stakeholders	Cooperation process between stakeholders. Knowledge about tasks and their planning

The return on experience documents about accidents, provide important information about the organization and implementation of in context fight actions. The set of general adequacy criteria, identified in the synthesis documents, has been extended with more specific criteria in relation to clearly defined situations.

In general, a lack of information about fight strategies as well as about justification of decisions is noticeable. This lack of information is a barrier to a full automation of the process of plan generation. This is not necessarily negative since GÉNÉPI is actually presented as a support tool for plan generation and not as a tool replacing the experts. During the generation of plans, GÉNÉPI requires a collaborative participation of the experts.

8.5.3 Linguistic Study of the Corpus

The application of the KOD method for conducting the linguistic study led to the development of Practical Models in accordance with the two following steps: (i) a terminological analysis of the corpus documents followed by (ii) a modelling operation by means of Taxemes, Actemes and Inferences.

The terminological analysis aims to identify in the corpus, terms and relationships used to describe the elements of the domain as well as their behaviour, considering the double point of view of the problem addressed and the method of resolution. The analysis consists in paraphrasing the sentences of the corpus documents to obtain simple sentences allowing to qualify the employed terms. The terms in question are representative of the three paradigms of the method. The result of this analysis is a terminological language where terms can be objects, values, relationships between objects and values, actions and inferences.

8.5.3.1 Taxeme Modelling

The modelling in the form of Taxemes consist in organizing terms representing objects and concepts of the triplet Td by means of binary predicates such as <Object, attribute, value>. The attribute defines a relationship between the object and a value. Five kinds of predicative relationships are defined: Classifying, Identifying, Descriptive, Structural and Situational.

The "classifying" relationship allows to represent that an object belongs to a class or a family of objects. The relationship is represented by the terms "Kind-of" or "is-a". The "Identifying" relationship characterizes an intrinsic property of an object. The attribute takes the value "is" (example: <cat, is, black>). The problem with this relationship is the ambiguity of the term "is" (exemple: <John, is, tall>). A better way is to define explicitly the intrinsic property. The "Descriptive" relationship characterizes explicitly intrinsic properties of objects. The attribute takes the value of the clarified property (Examples: <cat, Color, Black> and <John, Height, Tall>). The "Structural" relationship allows to model the constitution of objects. The attribute takes the value "compose" or "is-composed-of". The "situational" relationship allows to locate objects in a spatial or temporal reference. The attribute takes the value of the preposition (on, in, top, bottom, before, after, etc.).

Example *Original text:* "... On November 13, 2002, the Prestige oil tanker flying the Bahamian flag, sends an emergency message from the Finisterre Cape ..."

Paraphrases

1. The Prestige is an oil tanker
2. The Prestige flies the flag of the Bahamas
3. On November 13, the Prestige is located at the Finisterre Cape
4. On November 13, the Prestige sends an emergency message

Taxeme modelling from example:

Taxemes	Relations
<Prestige, IS A, oil tanker>	Classifying
<Prestige, FLAG, Bahamas>	Descriptive
<Prestige, LOCATION, Finisterre Cape>	Situational
<Prestige, DATE, November 13th>	Situational

The last paraphrase is related to an action, so it will be modelled by means of an acteme. The extent of this analysis at the Corpus, have allowed obtaining the set of taxemes needed for the representation of the universe described by the corpus of documents. An object of the real world is modelled by the sum of related taxemes.

8.5.3.2 Acteme Modelling

Obtaining actemes consists in identifying verbs of the corpus documents that represent the activities carried out by human or artificial operators. The modelling in the form of actemes consists in organizing terms within a 7-uplet structure that represents activities associated with elements of the triplet Td:

<Action manager, Action, Addressee, Properties, Stat1_Addr, Stat2_Addr, Instruments>

The action is performed by the Action manager by means of Instruments and it is applied onto the Addressee which undergoes a change of status (Stat1_Addr → Stat2_Addr). Each element of the 7-uplet structure is a taxeme and the action modifies at least one of the addressee attributes. The following example illustrates how to extract actemes from the corpus:

... The Prestige sends an emergency message...

The activity is "SENDING an emergency message". Once identified, the activity is translated into a 7-tuple (the acteme):

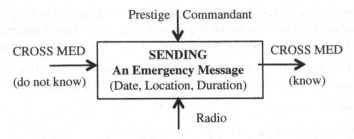

Action: FLUSHING	
Components	**Values**
Action Manager: Operator	{Human Means}
Addressee: Substratum	{Sand, Stone, Concrete, Rock, etc.}
Addressee State1	{Polluted, Cleaned}
Addressee State2	{Polluted, Cleaned}
Instruments	{Pump + Water_Hose + Recovery_Means}
Properties	Efficiency
Suitability Criteria	Viscosity Pollutant, Pollution level, Kind Of Substratum

Fig. 8.4 Two examples of actemes. One is represented by means of the datagram form (SENDING An Emergency Message) and the other (FLUSHING) by means of the table form

<Prestige Commandant, SENDING an emergency message, CROSS MED, (date, location, duration), CROSS MED (do not know), CROSS MED (know), Radio>

Where CROSS MED means "Centre Régional Opérationnel de Secours et de Sauvetage en Méditerranée", which is the French organism that receives any emergency messages from ships in difficulties. Figure 8.4 illustrates the different forms to represent actemes. For fight actions, it has been necessary to extend the formalism to take in account suitability criteria:

<Action Manager, Action, Addressee, Properties, Suitability Criteria, State1, State2, Instruments>

8.5.3.3 Inference Modelling

The modelling in the form of inferences consists in representing the elements of the corpus that characterize the cognitive activities of humans or machines.

Inferences are the basic elements of the Interpretation/Intention paradigm. An inference is the mental process, which consists in drawing a conclusion from a series of propositions accepted as true (premises).

In this study, the Interpretation addresses pollution situations and the Intention concerns fight action planning. Premise propositions are resulting from the interpretation of the situation elements. They are obtained from observation and therefore, they are held to be true. The conclusion is related to choose (or not) an action.

Let us consider the following example concerning the use of dispersant products:

> ... In general, dispersants should not be used in areas where water circulation is not good, close to spawning, coral reefs, shell deposits, wetlands and industrial water intakes...

From this extract, the following inferences have been produced:

- IF "spawning areas close" THEN "do not use dispersants"
- IF "coral reefs close" THEN "do not use dispersants"
- IF "shell deposits close" THEN "do not use dispersants"
- IF "industrial water intakes close" THEN "do not use dispersants"

Where "spawning areas close", "coral reefs close", "shell deposits close" and "industrial water intakes close" are the premise propositions. The observation (and interpretation) will give them the value True or False. To use dispersants, all the values have to be True. The suitability criteria associated to each fight action are the result of inference analyses.

8.5.4 Conceptualization

This step consists in developing the cognitive model from the practical models. At the end of this step, three sets of concepts are obtained: those associated with the objects and their properties, those associated with actions and those associated with the patterns of reasoning. The abstraction from practical models into a cognitive model is based on the operation of classification. This produces taxonomies, actinomies and patterns of reasoning.

8.5.4.1 Taxonomy Building

Taxonomy building is based on the analysis of taxemes. The first step consists in solving problems induced by homonym and synonym terms, with the objective to build a common terminology. The second step consists in analyzing the nature of attributes (or relationships) that characterize each object. From the nature of these attributes will depend the building of taxonomies (relationships "kind-of" or "is-a") or others kinds of tree structures (relationships "is-composed-of", "is-on", etc.).

As an example, from the analysis of the set of taxemes it was found that the term "Skimmer" is meaningful and thus it deserves the status of concept. It is significant of a set of recovery devices (modelled by means of taxemes). The definition of a concept is achieved by combining the whole knowledge about it. As a result of the analysis of the knowledge related to "Skimmer", the taxonomy in Fig. 8.5 has been built and the "Skimmer" concept is defined through its attributes as shown in Table 8.4.

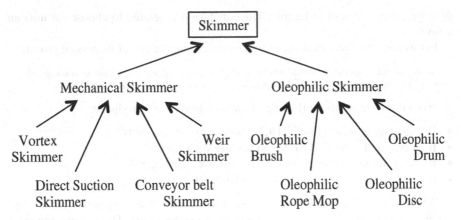

Fig. 8.5 The Skimmer taxonomy ("kind-of" relation)

Table 8.4 The attributes of the "Skimmer" concept

Skimmer	
Attributes	Values
Kind of	Hydrocarbon recovery devices
Owner	{Customs, National Marine, …, Civil Security}
Storage town	{Toulon, Port de Bouc, Marseille}
Storage location	{Name of the location in the town}
Quantity	Integer
Weight	Real
Flow	Real
Selectivity	{Poor, Good enough, Good, Very good}
Recovery rate	Real
Debris (usage limit)	{Very sensitive, Sensitive, Not sensitive}
State of the sea (usage limit)	{0, 1, 2, 3, 4}
Pollutant viscosity (usage limit)	{Fluid, Not viscous, …, Highly viscous}

8.5.4.2 Acteme Abstraction

One result from acteme analysis is that actemes are organized into five main action categories:

- Actions related to pollutant behaviour,
- Actions related to the behaviour of the stricken ship,
- Actions related to reasoning patterns,
- Actions related to CLARA 2 services.
- Actions related to operations against pollution,

Amongst actions related to pollutant behaviour we can cite: "Evaporation" and "Dissolution". Amongst actions related to stricken ship behaviour, we can cite:

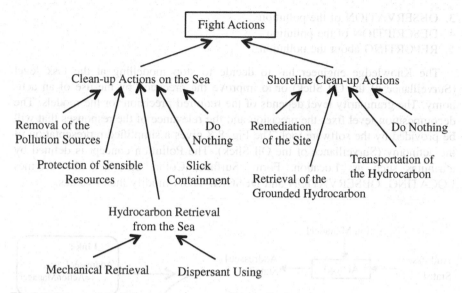

Fig. 8.6 Extract of the fight action taxonomy ("kind-of" relations)

"Listing to starboard" and "Sending an emergency message". The actions related to reasoning patterns such as "Choosing the shoreline clean-up methods" are used to select or to plan fight actions. To be performed, they use the suitability criteria associated to each acteme. The actions that belong to the CLARA 2 services category are implemented to improve the GENEPI functionalities.

The actions of the last category are fight actions (Fig. 8.6). They are divided into two main classes: (i) the shoreline clean-up methods and (ii) the clean-up methods on the sea. Some actemes of the fight action category can be organized in a structural and temporal way to form actinomies. The interest of this kind of structure is that actions are already planned.

8.5.4.3 Actinomy Building

Actinomies are obtained by acteme association in order to provide a behaviour description of the domain objects or to model processes associated with the Problem Solving Method. The structure of the actinomy is defined by links between actèmes. The kinds of links are (Addressee, Action Manager), (Adressee, Adressee), (Action Manager, Adressee) or (Action Manager, Action Manager). The Fig. 8.7 gives a generic representation of an actinomy structure.

The following example illustrates the process that has to be followed by the staff of reconnaissance aircrafts for the pollution surveillance. The "Surveillance of the Oil Slick" is a composite activity that is composed of five basic actions (actemes):

1. PREPARING a flight plan
2. LOCATING the pollution

3. OBSERVATION of the pollution
4. DESCRIPTION of the pollution
5. REPORTING about the pollution

The Knowledge engineer has to decide to stop modelling at the task level (Surveillance of the Oil Slick) or to improve the precision by the use of an acti-nomy. The granularity level depends of the required precision for the models. The decomposition level fixes the precision and the relevance of the responses that will be provided by the software tool. The Fig. 8.8 gives a simplified representation of the actinomy (Surveillance of the Oil Slick). The Pollution concept is defined by attributes such as: Location, Form, Surface, Color, etc., and the actemes LOCATING, OBSERVATION and DESCRIPTION modify theirs values.

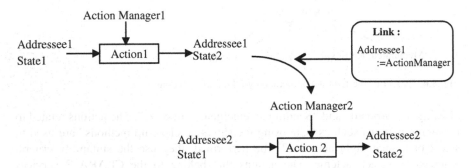

Fig. 8.7 Generic representation of an Actinomy

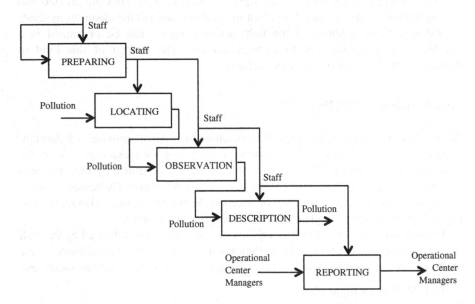

Fig. 8.8 Simplified representation of the "Oil Slick Surveillance" actinomy

8.5.4.4 Building of Reasoning Patterns

In our case study, we have not established reasoning patterns made up of logical links between inferences to produce a complex reasoning. The reasoning Template implemented is that of the inference rule:

Premise Proposition 1.

Premise Proposition 2.

…

Premise Proposition n

Conclusion

As it has already been discussed in the section on inferences (Sect. 5.3.3), premises propositions are related to observed situations and the suitability criteria are extracted from these premise propositions. Thus, inference rules for selecting or not an intervention action are of the form:

$$c_1{}^\wedge c_2{}^\wedge \ldots {}^\wedge c_n \rightarrow \text{True/False}$$

Where $c_1, c_2, \ldots c_n$, are the suitability criteria associated to each fight action and the symbol ($\wedge$) is the logical AND operation. The conclusion of the rule is whether the action can be selected or not. A suitability criterion is satisfied if its observed value in the real situation is compatible with the constraints imposed by the implementation of the action. All criteria must be met to select the action characterized by the inference rule.

8.5.5 Formalization and Software Model

Ontology formalization (the cognitive or conceptual model) consists in selecting or building the formal language capable of integrating all the properties of the ontology. But, it must also be relevant to the use that have to be made of the ontology (the operational dimension).

For the formal representation of the GENEPI ontology, the frame-based representation language of the Protégé platform has been chosen. The two reasons for this choice are:

1. The representation is very close to the conceptual model we have reached by the application of the KOD method. So the formal model can be read as easily as the conceptual model.
2. The Protégé platform is a work environment widely used, efficient, open source and providing an Application Programming Interface (API) required to develop the GENEPI functions.

As an example, the Fig. 8.9 presents the case of the concept "Skimmer" ("Récupérateur" in French), expressed in the frame-based representation language as proposed by Protégé. In the Fig. 8.9a (the top picture), the taxonomy can be

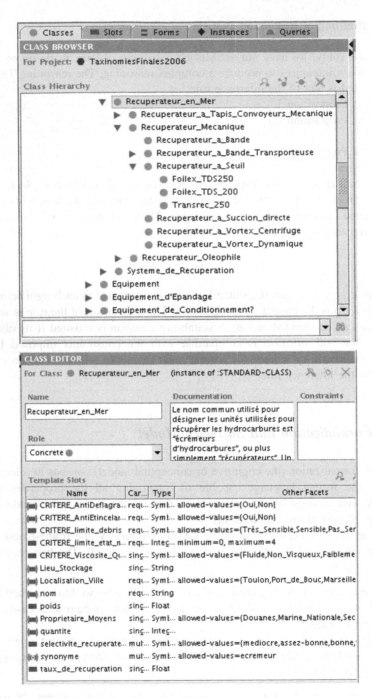

Fig. 8.9 Formalization of the ontology using a frame-based representation language in the work environment offered by Protégé. The "Skimmer" taxonomy (at the top picture) and the definition of the Skimmer concept (at the bottom picture)

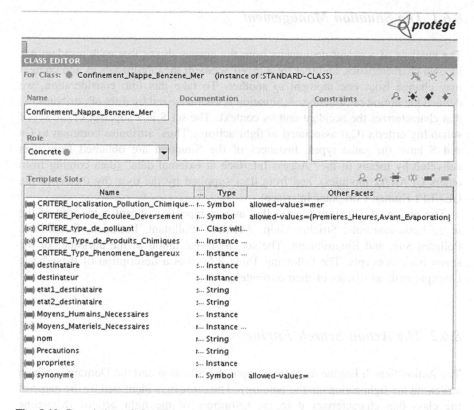

Fig. 8.10 Description of the "Benzene Confinement" fight action ("Confinement_Nappe_Benzene_Mer" in French)

easily read where the hierarchy levels of the tree structure (Fig. 8.5) are represented here by indentations. In the Fig. 8.9b (the bottom picture), the concept "Skimmer" (Table 8.4) is presented where attributes are supplemented by the suitability criteria expressed in the form: CRITERE_xxx.

To illustrate the representation of fight actions, the Fig. 8.10 presents the case of the benzene confinement. The acteme components are modelled as well as the suitability criteria.

8.6 Architecture of the GENEPI Module

The architecture of the GENEPI module (Fig. 8.11) has been designed around the ontology enriched with the instances of the concrete classes. The association of the ontology with instances constitutes the knowledge base of GENEPI.

8.6.1 The Situation Management

The analysis of accident documents from the corpus shows that each accident has its own characteristics, and that for a given accident the circumstances and context may change from one moment to another. To take this into consideration, we defined the notion of Situation. A Situation is characterized by a set of attributes (S) that characterizes the accident and its context. The set S is a superset of the set of suitability criteria (Ca) associated to fight actions. Thus, attributes common to Ca and S have the same types. Instances of the Situation are obtained from data delivered by means of the "Access Interface at external data" (data coming from others CLARA2 modules), and from data supplied by the user (by means of the GENEPI Human Computer Interface).

The notion of Situation is modelled as a composite concept composed of the seven basic concepts: Stricken Ship, Accident, Pollutant, Pollution, Conditions, Polluted Site, and Environment. The set (S) is the union of the attributes of the seven basic concepts. The following Table 8.5 gives a description of these basic concepts with an extract of their attributes.

8.6.2 The Action Search Engine

The Action Search Engine receives as input the Situation and the Domain in which searching the fight actions in the ontology. The domain is identified by the name of the class that characterizes it in the taxonomy of the fight actions (Shoreline

Table 8.5 Description of the seven basic concepts that compose the situation concept

Concepts	Definition	Example of attributes
Stricken ship	Ship affected by accident	Name, kind of ship, cargo, damage, etc.
Accident	Event leading to damages to the environment	Kind of accident, severity, location, date, etc.
Pollutant	Hydrocarbons or chemical products that have undesired effect onto the environment	Color, density, viscosity, biodegradability, etc.
Pollution	Substances resulting from Pollutant actions onto the environment	Location, quantity, concentration, viscosity, form, etc.
Conditions	Climatic and oceanographic conditions	Sea temperature, Sea state, wind speed, etc.
Polluted site	Coastal place covered by pollution	Pollution level, topography, coastal substrate, accessibility, etc.
Environment	Is covering ecosystems and economical activities	Kind of Ecosystem, vulnerability, activity

Clean-up Actions, Clean up Actions on the sea, etc.) (Fig. 8.6). The domain is provided by the user. As a result, the Action Search Engine provides four sets of fight actions:

- The set A, which contains the actions where all criteria are verified,
- The set B, which contains the actions where at least one of the criteria can not be assessed by lack of information in the situation,
- The set C, which contains the actions of which at least one criterion was not satisfied,
- The set D, which contains the actions of the set B enriched by criteria not assessed.

Rules for selecting fight actions are based on suitability criteria and values taken by the corresponding attributes of the situation. Rules are of the form:

$$c_1 {}^{\wedge} c_2 {}^{\wedge} \ldots {}^{\wedge} c_n \rightarrow \text{True/False}$$

With $c_1, c_2, \ldots c_n$, the criteria associated to a fight action. The conclusion of the rule is about the possibility whether or not to select the action. A criterion is satisfied if the value taken by the corresponding attribute of the situation is compatible with the criterion constraints.

Upon the receipt of the Situation, the action-selecting algorithm analyses the actemes involved in the Search Domain. From each acteme, it extracts the criteria and it applies the selection rules previously presented. According to the results obtained, the acteme is placed in the corresponding set (A, B, C or D). After having run the algorithm, if the user is not satisfied with the result, he can enrich the initial situation to assess the criteria that could not be assessed previously. This new run will reduce the size of the set B, by transferring actions in the set A or in the set C. The algorithm is independent of changes in the ontology.

8.6.3 The Plan Generator

Fight action plans are the result of a collaborative work between GENEPI and the user. From the set A (set of actions where all criteria are satisfied), the user selects actions to constitute the Plan. Once the actions are selected, the Plan Generator produces a document where every action is completely defined. The user then has at his disposal, for each action, the following information:

- A detailed description of the fight action,
- A detailed description of human and material means required for its implementation,
- A detailed description of precautions and safety measures to be followed for its implementation,
- A reminder of the suitability criteria.

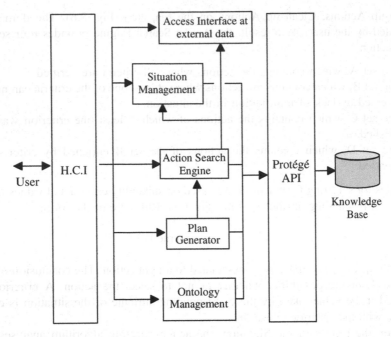

Fig. 8.11 Architecture of the GENEPI module

Material means are also subject to a precise description that facilitates their Mobilization.

8.6.4 The Ontology Management Module

This module provides users with the functions needed for maintenance (updating, adding and deleting classes, attributes and instances) and consultation (searching knowledge) of the ontology.

8.7 Conclusion

The purpose of this chapter is to demonstrate that cognitive approaches offer very powerful engineering environments to tackle the issues raised by risk management. The domain of the responses concerns the design of intelligent machines to support expert reasoning, and more precisely to design support systems for problem-solving and decision-making.

The investigated issue is the planning of actions to fight accidental marine pollutions. The production of fight action plans is a complex process implemented in operational centers under strong time constraints, climate constraints and environmental constraints. These operational centers are constituted of stakeholders having diverse cultural and professional backgrounds and, which may have different goals in how to respond to accidental situations.

The response proposed is a software tool (GENEPI within the CLARA 2 project) to support stakeholders to plan fight actions during emergency or critical situations. The plans generated by GENEPI take into account accidental situations and their evolution over time.

The methodological process to design GENEPI is based on ontology elaboration. The purpose of that ontology is to represent a common way of working within operational centers that is recognized and accepted by all stakeholders. The resulting ontology structures and models the domain (maritime accidents causing marine pollutions) according to the problem to solve (assisting stakeholders to produce fight action plans) and to the problem solving method (a cooperative software tool).

The ontology is obtained through a cognitive approach, which consists in applying the Knowledge Oriented Design method (KOD). KOD has proven to be adequate (i) to acquire knowledge from a corpus of documents; (ii) to conceptualize this knowledge; and (iii) to develop ontologies at the knowledge level. With KOD, we can actually build a corpus, identify the terminology used in a domain and develop the conceptual model (ontology). KOD covers the cycle that is generally accepted for the construction of ontologies, especially phases of the linguistic study and of the conceptualization.

KOD is a relevant method for building ontologies. Based on research carried out in cognitive anthropology and linguistics, it suggests a methodological framework for the collection and organization of knowledge, depending not only on the domain but also on the problem to solve and how to solve it. In addition, if some knowledge is missing, KOD can guide the work of the knowledge engineer to identify it, if necessary. The main capabilities required by methods for building ontologies are: (i) language capabilities; (ii) conceptualization capabilities; (iii) capabilities to model reasoning; (iv) capabilities to express the generic features of concepts; and (v) the consensual features of concepts.

References

1. Uschold, M. (1998). Knowledge level modelling: Concepts and terminology. *The Knowledge Engineering Review, 13*(1), 5–29.
2. Uschold, M., & Grüninger, M. (1996). Ontologies: Principles, methods and applications. *Knowledge Engineering Review, 11*(2), 93–136.
3. Vogel C. (1988). *Génie cognitive*. Paris: Masson (Sciences cognitives).
4. Rempec. (2002). *Guide pour la lutte contre la pollution marine accidentelle en Méditerranée, Partie D, Fascicule 1*. Avril 2002.

5. CLARA 2 Consortium: École des Mines d'Alès, le Cèdre, IFREMER, Météo France, IRSN, TOTAL, EADS, Géocéan, UBO, INERIS, SDIS 30, Préfecture Maritime de la Méditerranée, le CEPPOL, LSIS. Projet ANR, 2006–2010.
6. Oxford Dictionary, online version. (2008, January 1st) http://www.askoxford.com.
7. ISCID. (2008, January 1st). Encyclopedia of science and philosophy, International Society for Complexity, Information and Science. http://www.iscid.org/encyclopedia/.
8. Wikipedia dictionary (http://en.wikipedia.org/wiki/Ontology).
9. Gruber, T. R. (1993). *Toward principles for the design of ontologies used for knowledge sharing*. Technical Report KSL-93-04, Knowledge Systems Laboratory, Stanford University. Revised in August 1993.
10. Gruber, T. R. (1995). Towards principles for the design of ontologies used for knowledge sharing. *International Journal of Human-Computer Studies, 43*(5/6), 907–928.
11. Bachimont, B. (2001). Modélisation linguistique et modélisation logique des ontologies: l'apport de l'ontologie formelle. In *Actes de la conférence francophone Ingénierie des connaissances* (pp. 349–368). 25–28 juin 2001, Plate-forme AFIA, Grenoble.
12. Maedche, A. (2002). *Ontology Learning for the Semantic Web*. Boston: Kluwer Academic Publishers.
13. Guarino, N. (1995). Formal ontology, conceptual analysis and knowledge representation. *International Journal of Human-Computer Studies, 43*(5–6), 625–640.
14. Guarino, N., & Giaretta P. (1995). Ontologies and knowledge bases: Towards a terminological clarification. In N. J. I. Mars (Ed.), *Towards Very Large Knowledge Bases*. IOS Press.
15. Fernández-López, M. (1999). Overview of methodologies for building ontologies. In Proceedings IJCAI-99, Workshop on Ontologies and Problem-Solving Methods (KRR5) (pp. 4.1–4.13). Stockhom, Sweden, August 2.
16. Pinto, H. S., & Martins, J. P. (2004). Ontologies: How can they be built? *Knowledge and Information Systems, 6*(4), 441–464.
17. Corcho, O., Fernández-López, M., & Gómez-Pérez, A. (2003). Methodologies, tools and languages for building ontologies: Where is their meeting point? *Data & Knowledge Engineering, 46*(1), 41–64.
18. Fernández-López, M., & Gómez-Pérez, A. (2002). Overview and analysis of methodologies for building ontologies. *The Knowledge Engineering Review, 17*(2), 129–156.
19. Grüninger M., & Fox M.S. (1995). Methodology for the design and evaluation of ontologies. In *Workshop on Basic Ontological Issues in Knowledge Sharing, IJCAI-95*. Montréal (Canada).
20. Uschold, M. (1996). Building ontologies: Towards a unified methodology. In *Proceedings of Expert Systems 96*. December 16–18, Cambridge. http://citeseer.ist.psu.edu/uschold96 building.html.
21. Uschold, M., & King, M. (1995). Towards a methodology for building ontologies. In *Proceedings of the IJCAI-95, Worshop on Basic Ontological Issues in Knowledge Sharing*. Montréal, Canada.
22. Fernández, M., Gómez-Pérez, A., and Juristo, N. (1997). METHONTOLOGY: From ontological art towards ontological engineering. In *Proceedings AAAI-97 Spring Symposium Series, Workshop on ontological engineering* (pp. 33–40). Stanford, California.
23. Aussenac-Gilles, N., Biébow, B. and Szulman S. (2000). Revisiting ontology design: a method based on corpus analysis. In R. Dieng & O. Corby (Eds.), *Proceedings of the 12th International Conference on Knowledge Engineering and Knowledge Management* (pp. 172–188). LNAI 1937, Springer.
24. Blazquez, M., Fernandez-lopez, M., Garcia-Pinar, J.M., and Gomez-Pérez, A. (1998). Building ontologies at the knowledge level using the ontology design environment. In *Proceedings of the Workshop on Knowledge Acquisition, Modelling and Management (KAW'98)*. Banff, Canada.
25. Gandon, F. (2002). Ingénierie d'ontologie: une synthèse et un retour d'expérience. Rapport de Recherche n° 4396, INRIA Sophia-Antipolis.
26. Dahlgren, K. (1995). A linguistic ontology. *International Journal of Human-Computer Studies, 43*(5), 809–818.

27. Dehainsala, H., Pierra, G., Bellatreche, L. and Aît Aimeur, Y. (2007). Conception de bases de données à partir d'ontologies du domaine : application aux bases de données du domaine technique. In F. Gargouri, D. Benslimane and P. Bourque (Ed.), *Les ontologies:mythes, réalités et perspectives* (pp. 215–230). 1ère Édition des Journées francophones sur les ontologies. Centre de Publication Universitaire.

28. Mercantini, J.-M., Loschmann, R., & Chouraqui, E. (2000). A provisional analysis method on safety of an urban industrial site. *People and Work Research Report (Safety in the Modern Society), 33*, 105–109.

29. Mercantini, J.M., Capus, L., Chouraqui, E., & Tourigny, N. (2003). Knowledge engineering contributions in traffic road accident analysis. In Ravi K. Jain, Ajith Abraham, Colette Faucher, Berend Jan Van der Zwaag (Ed.), *Innovations in Knowledge Engineering* (pp. 211–244).

30. Mercantini, J.-M., Turnell, M.F.Q.V, Guerrero, C.V.S, Chouraqui, E., Vieira, F.A.Q et Pereira M.R.B. (2004). Human centred modelling of incident scenarios. In *IEEE SMC 2004, Proceedings of the International Conference on Systems, Man & Cybernetics* (pp. 893–898). The Hague, The Netherlands, October 10–13.

31. Caroll, J. M. (1997). Scenario-Based Design. In M. Helander, T. K. Landauer, & P. Prabhu (Eds.), *Handbook of Human-Computer Interaction (Chapter 17)*. Elsevier Science B.V: Second completely revised edition.

Chapter 9
The Operators' System of Instruments: A Risk Management Tool

Grégory Munoz, Christine Vidal-Gomel and Gaëtan Bourmaud

Abstract Previous analyses of the working activity of professionals in charge of safety in industrial companies, also called preventionists, have shown that the purpose of this activity consists of establishing a process of "pragmatization of regulations." This is an adaptation of the regulations, relative to processes from the texts of law of general order towards their implementation in a context [1]. We have analyzed these processes according to the instrumental approach of Rabardel [2]. In this perspective, we focus on the safety workers' systems of instruments [3–8]. These resources, developed according to the workers' experience, render their activity more reliable. They are also of heterogeneous character: simultaneously material, symbolic or of cognitive order. The systems of instruments present certain properties, in particular being structured according to the experience and skills of the workers, but also characterized by the complementarities and redundancies of their functions, following the example of a security system; this takes into account the elements of the context, the constraints and the resources of the activity [3]. In the following case study, we present the system of instruments of a preventionist, in which the regulations function as a "pivot instrument" of this system. It is from the regulations that the worker establishes the diagnosis of the safety level of his or her company and develops his or her preventive and formative actions [9]. To conclude, we develop a discussion about the design of the preventionists' training.

G. Munoz (✉) · C. Vidal-Gomel
Centre de Recherche en Éducation de Nantes
(CREN- EA 2661), Université de Nantes (Nates University),
BP 81227 Chemin de La Censive du Tertre, 44312 Nantes Cedex 3, France
e-mail: gregory.munoz@univ-nantes.fr

C. Vidal-Gomel
e-mail: vidal@univ-nantes.fr

G. Bourmaud
Centre de Recherche sur le Travail et le Développement (CRTD),
Conservatoire National des Arts et Métiers (CNAM), 41 rue Gay-Lussac,
75005 Paris, France
e-mail: gaetan.bourmaud@cnam.fr

© Springer-Verlag Berlin Heidelberg 2015
J.-M. Mercantini and C. Faucher (eds.), *Risk and Cognition*,
Intelligent Systems Reference Library 80, DOI 10.1007/978-3-662-45704-7_9

9.1 Introduction: Instruments and Risk Management

The art of managing risks would tend to become a political art, while it formerly came from an essentially technician realm. (Viet and Ruffat 1999, p. 3) [10].[1]

Accident prevention in at-risk systems was first oriented by retroactive methodologies, based on accident and incident analyses. During the 1990s, technological evolutions and the transformations of work organizations lead to the development of proactive approaches [11]. The concept of resilience should be understood in this perspective. It was composed with the aim to better anticipate undesirable events [12]. In this framework, the centre of the analysis evolves, from deviations and abnormal situations, to the normal functioning of socio-technical systems and their capacity to face unexpected events. Then, resilience is the "intrinsic capacity of an organization or a complex socio-technical system to maintain or recover a stable state, which allows it to fulfill operations (e.g. production, etc.) after a major event and/or permanent stress" [13]. This approach is coherent with the French ergonomics orientation,[2] which, according to Faverge [15], stresses that operators are actors of system reliability, particularly in regulation activities.

In this perspective, the analysis of the effective practices of operators in charge of safety (preventionists) is inescapable. They are key actors of risk management in a company. In a previous study [1], we observed that their work consists of:

1. Constituting a reference of what is deemed to be "a secure company,"
2. Diagnosing the state of safety of their company by comparing it with regulation requirements,
3. Planning and carrying out preventive or corrective actions.

To achieve their assignments, regulations constitute the operators' main instrument [16]. Such an assertion seems commonplace. However, it is a question of considering whether it gives rise to a process of appropriation [7], which underlines some questions about the relationship between the operators and the safety rules.

Rules are operational principles, established to reach a safety objective for material equipment, operators and/or the environment [17]. They constitute a legal reference, which allows one to establish, to some degree of precision, who or what is responsible for an accident [18]. The operators' relations to these safety rules are primarily analyzed in terms of compliance or violation. Violations are defined as deliberate but are carried out without the intention of causing damage or harm [19]. In some cases, the operators do not have sufficient knowledge of the rules or the consequences of non-compliance; they take a risk [20]. Following the same reasoning, one would consider that the safety rules are the one best way to manage risks. Safety rules are developed by experts on the basis of task and risk analyses. The objective of these experts is to supervise, control human behavior and avoid errors.

---

[1] The translation is from the authors of the chapter.

[2] See [14] for details.

Based on research realized in complex and high-risk companies, another point of view recognizes that safety rules cannot be complete with regard to the diversity, variability and complexity of real situations [21]. The experience and expertise of the operators is necessary to adapt safety rules to the specificity of situations, as well as to unforeseen circumstances. In this perspective, the rules must not be simply applied, and violations are not only understood as taking risks [21]. This point of view takes into account the operators' contribution to the resilience of sociotechnical systems. In the same angle, we consider that safety rules, their appropriation by the operators and their use relies on the development of occupational competencies when the safety functions of the rules are reached.

For instance, Mayen and Savoyant [22], who investigated the understanding and uses of safety rules, identified a process of rule "reinvention," meaning the development and attribution of a new meaning to the rules, which transforms them into necessities. The rules then become recognized and reinvented in their logical necessity, with regard to a global coherence of the operators' activity and the safety system. Understanding the safety functions of a rule is central in the development of skills. We propose to consider that this process is linked to the elaboration of new instruments: resources of the operators' activity in a situation.

To be more precise, an instrument "is not only a component of the world exterior to the subject, available to be associated with the action", but it is also a "construction, a production of the subject" ([2], p. 118). An instrument is an "artifact in a situation, incorporated into a specific use" (Ibid, p. 116); therefore, "the instrument is not one per se but the result of an association of the artifact with the action of the subject" (Ibid, p. 79). An artifact, from an anthropological standpoint, is considered to be any manmade object; it can be material or symbolic but appears exterior to the subject and can be used by the subject. An instrument is a mixed entity, containing an artifactual component and a structural component: the subject's schemes[3] of use.

In analyzing the different uses of an artifact, catachresis must be taken into account.[4] Catachresis is a characteristic of instrumental genesis, i.e. "the expression of a specific activity of the subject: the production of instruments and, more generally, his or her means of action" ([2], p. 12). Traditionally in the French ergonomics orientation, they are understood as unplanned uses of an artifact [24]. With the instrumental genesis framework, we understand this as a process of skill development. These ideas are not new and encounter the lineage of activity theories [25] to show that the instruments are not given to the users but are the product of their personal construction [26–28].

Instruments are not isolated from each other or independent, they form a system: an organization in a coherent set [2]. It is a question of considering systems of instruments and investigating the relations developed between different instruments, which compose an identical system [3–6].

---

[3] Schemes are invariant organizations of action for a class of situations [23].

[4] Catachresis: "the use of a tool on the place of another one or a use for which it is not designed" ([2], p. 11).

Numerous studies are based more or less on the instrumental approach of Rabardel [2], which allows one to understand how subjects transform artifacts into instruments (instrumental genesis). However, little research has attempted to consider the operators' systems of instruments implemented in work situations [3–8, 29–31].

The main objective of this chapter is to identify some characteristics of such a system in the preventionists' work activity in industrial companies. Characterizing these systems formulated by operators, their safety functions and their limits could be key in designing training programs. We present here a study carried out via several interviews with preventionists. The interviews have been transcribed and analyzed to identify any traces of the system of instruments.

Having clarified the analysis of safety rules in terms of instruments implemented in professional situations, we now present the methodological framework of our case study, and we examine a preventionist's system of instruments. Preventionist activity requires a process of "pragmatization" of safety rules. The regulations constitute a "pivot instrument" within their system, allowing the subjects to perform their main assignments. We will present a functional analysis of the different instruments in the system. To conclude, we shall discuss the consequences of our initial results, pertaining to the design of training programs for preventionists, and also for risk management in a more general perspective.

9.2 Systems of Instruments and Safety Rules

9.2.1 Characteristics of the Systems of Instruments

Bourmaud has highlighted the main characteristics of systems of instruments in examining previous research, using the following concept [2, 7, 8, 30–32]: the different components are heterogeneous, the functions are complementary and redundant, a specific instrument is the "pivot" of the system, and these systems are robust and adaptable [3–6].

9.2.1.1 The Heterogeneousness of Instruments of the System

Formal and institutional instruments, for example safety rules, coexist with unofficial ones, for example the non-formal use of a safety rule. The resources participating in the systematic organization of instruments are heterogeneous in nature.

9.2.1.2 The Pivot Instrument of the System

Among all instruments composing the system, one is quite exceptional: the pivot instrument. Different indicators can be used to identify it.

9.2.1.3 The Complementarity and Redundancy of Functions of the System

Systems of instruments present the double characteristic of complementarity and redundancy of functions. For example, two different instruments may have different and complementary functions for managing risks, and the two instruments may reach the same safety functions. In this case, only one of them is systematically chosen, according to the situation's characteristics and to availability and accessibility.

9.2.1.4 The Robustness and the Adaptability of the Systems of Instruments

Finally, the double characteristic of complementarity and redundancy of the instrument's functions contributes simultaneously to the robustness of the system, its flexibility and its adaptability to face the variability and diversity of situations.

9.2.2 System of Instruments and Safety Rules

In the maintenance of electrical systems, Vidal-Gomel [7, 8] brings to light a system of instruments developed by the electricians. A safety rule that is analyzed as an artifact gives rise to several instruments. The rule in question here concerns how to ensure that the operation has been effectively shut down, by verifying that the power in close proximity to the workstation has been cut. This operation is a major means of working safely. Particularly in a situation with a latent connection error [19]. In these situations when operators turn off the electric current, it continues to flow; the operators are not always able to detect it. These are critical situations for novices but also for experienced operators [33]. Different uses of the same safety rule may be implemented to identify this type of situation: (1) checking the connection node upstream of the element, which is used for cutting the power (formal use of the safety rule), (2) verifying both connection nodes of this element (upstream and downstream). This allows the detection of different connection errors that could exist on the concerned circuit. It therefore constitutes a more precise instrument than the formal application of the safety rule.

These two instruments ensure that the safety goal (switching off the system) is reached. Other means include localized power control, which is useful for checking the absence of a connection error at a confined section of the electric circuit:

- Checking the downstream connection node of the circuit breaker. This control composes the third instrument, which is developed according to the same artifact, that is to say the safety rule.

- Using the circuit breaker's test button. This operation automatically lowers the lever of the circuit breaker and switches it off. With this operation, the operator verifies that the circuit breaker[5] functions properly and that there is no connection error at this point.
- Checking that the neon lighting has been switched off with the corresponding circuit breaker.

These three instruments partly meet the safety functions of the rule. They are complementary. Their joint use is a factor in the reliability that the system has been powered off, a crucial task for risk management in electrical maintenance.

Our approach to safety rules aims to go beyond an analysis in terms of compliance or violation of the rules. It also takes into account the development of different operators' resources according to their experience. More precisely, the operators' systems of instruments differ according to their experience in a specific domain of electrical maintenance: in the trade and in the company [8]. These systems are composed of heterogeneous resources: simultaneously material (the lever of the circuit breaker), symbolic (the safety rule) or semiotic (the neon lighting).

9.2.3 System of Instruments and Resilience

The results obtained in another work domain with the Method of Failure and Substitution of Resources[6] (MFSR) allow to specify the characteristics of the systems of instruments [3].

Here, we will develop the idea that the MFSR is useful in the analysis of reliability and adaptability in work systems. It stresses that the double characteristic of function complementarity and redundancy contributes simultaneously to the robustness of the system and to its flexibility and adaptability when facing situations of variability and diversity.

Indeed, the existence of other artifacts in a system shows that a failing instrument can be easily replaced by another which fulfills equivalent functions, since it is another usual artifact of the operator. Furthermore, in some cases, several different artifacts (substitution resources) ensure function redundancy, and they can be used according to situational characteristics. For example, checking that neon lights have been switched off is not always pertinent. In this case, operators can use the test button, the control on the downstream node of the circuit breaker, and check that both connection nodes of the element are cut.

In this way, one or several alternative solutions are available in the case of failure of a resource, or according to the functions of a resource. Mostly, substitution

---

[5] The test button generates an electric fault that must be automatically detected by the circuit breaker and then the cut-off must be achieved.

[6] Here "resource" is synonymous to instrument.

resources are usual artifacts and are intrinsically part of the systems of instruments, and they allow an operator to respond to possible failures. The substitution and the substitution conditions—two dimensions of substitution resources—sharply stress:

- The substitution resource is usually judged as less "effective, practical, safe, precise, etc." but it is an available possibility.
- In the majority of substitution cases, no specific substitution conditions exist; the substitution resource is already a component of the system of instruments.

Our case study attempts to address these questions by exploring the preventionists' system of instruments and especially by focusing on resilience proprieties, such as substitution and robustness.

9.3 Methodology of the Case Study

9.3.1 Collected Data: A Study of the "Redefined Task"

The aim of this case study is to consider any characteristics of the systems of instruments. We have adopted a qualitative case study approach [34]. The data examined here was obtained during a series of three interviews with nine preventionists, each working in an industrial environment. The subjects were considered to be experts, since they could have also been either trainers or tutors for learners in vocational training centers. We also carried out daily work observations and participated in safety clubs.

In order to identify the systems of instruments developed by preventionists, we must try to understand their "redefined task" [35]. The task defined from the point of view of the subject comprised the operator's representations of his or her work, the way it is realized, his or her personal values, etc. The redefined task differs from the prescribed task in that the task is defined from an organizational point of view, including the task defined by the individual who realizes it and the task that is actually accomplished. Our process of data collection on the basis of interviews was organized in three phases: the subjects' definition of the work situations, the validation of these definitions and the confrontation with their various points of view. In the first interview phase, the subjects were asked to explain what they consider to be a difficult situation in their daily work. The transcribed interviews were divided into themes and sub-themes and validated during a second interview. The themes pertained to the theoretical contents of safety; the sub-themes were related to episodes corresponding to real work experience. At this moment, we provided the subject with an initial proposed categorization. The aim of this second phase was to specify or further explicate the different points of their discourse. In the third phase, we organized a confrontation with other operators: other preventionists were asked to comment on an anonymous, transcribed interview. The transcription therefore became a document used for interviewing a group of operators.

The data mobilized in this chapter concerns the first two phases of our interview process with an operator who is also trainer in a vocational training centre in the industrial field. In the same vein as the work of Creswell [34], we chose to present this in-depth portrait because this particular professional explicitly develops his activity during a long, 3-h interview, completed with another, lasting 1 h. We will focus here on the theme concerning the tools used.

9.3.2 The Subject's Characteristics

After obtaining a high school diploma and achieving a 2 year university degree in science, followed by a 2 year technical degree in chemistry, this operator (we will call him Subject A) became an engineer at the *Ecole Polytechnique* at the University of Grenoble. He specialized in the field of hygiene, safety and the environment. During his career, he handled fire management and the implementation of a safety management system in a company manufacturing industrial ink-jet printers. In another job, he dealt with machine conformity as a preventionist. His studies in ergonomics supplied him with a constant concern for the human being at work, which moreover, is highly visible in his comments.

9.3.3 Data Analysis: Components of the System of Instruments and Functional Analysis

The data analysis here consists of two general phases: after having determined the components of the operator's instrument system, we carried out a functional analysis. To highlight the components, we illustrated their specificities with extracts from the interviews with the professional. For system 1, we considered all artifacts mobilized by the professional and, for each one, we determined the functions and goals fulfilled.

For instance, Table 9.1 synthesizes the following explanations from the operator[7]:

37. **Interviewer**: What did you, when you arrived at your company or during your initial training courses, what did you use every day as tools? If you had books, if you had?
38. **Subject A**: Ah okay, yes there is the **Labor code, it is the most important tool**.
39. **Interviewer**: Because I don't put the documents …

---

[7] Translated from French.

Table 9.1 Example of the functional analysis of instruments

No	Artifacts	Functions	Meta-goals	Comments
38	Labor code	F1. to be capable of considering the statutory risks F2. to be capable of considering the risks of safety	1. to Establish the expected reference of a "secure company"	"Fundamental tool" (38); "basic tool" (40)

40. **Subject A**: The **Labor Code, the basic tool**, more than the Labor Code, for me what I use is the **Permanent Dictionary of Safety and Working Conditions**, because in fact, it's the **interpreted Labor Code**. It is not simply the texts of laws, it is a little, it goes a little further, you don't have that in the Labor Code, and you have a **certain number of orders, decrees, and European directives**. So **that allows you to simply have more information**. That it is **the work tool**, yes, it is the **basic work tool**, it's true, I hadn't thought about that.

Afterwards, we attempted to collect other characteristics of the artifact. For example:

57. **Subject A**: (…). I shall say, that's right in fact, there is theoretical knowledge; it is **the Labor code that is the theoretical knowledge**. After that, there is also technical knowledge, for example to know how to use certain… if among the tools there is in particular a "causal tree", to go back a little, to know, when we had the accident, to be capable of starting a verification of the accident, to set up actions, it will be passive actions or corrective prevention, as you want. We had the accident, we tried to set up actions to avoid it happening again. That's a tool, it's sure that there is certain number of tools. You can't arrive in a company and improvise like that because you have to know how to use them.

9.4 Results: The System of Instruments of a Preventionist

Firstly, we will present some general results pertaining to preventionists' work. This will compose the general context of our analysis of A's system of instruments.

9.4.1 General Results of Work Analysis

Preventionists are in charge of complying with safety regulations and implementing them through adaptation [1, 9, 36]. This means they have to conceptualize the interest and legitimacy of the rules. In doing so, they carry out a process of "pragmatization of regulations". This process is an instrumental genesis [2] that transforms the formal rules (artifacts) into possible instruments for action [7, 8] by other actors. The concept of "pragmatization." corresponds to the process by which rules are adapted and transformed from their promulgated forms in the texts of regulations. These texts are generic and able to be implemented in all work situations; they are specific to each workstation. The concept also concerns, from a psychological standpoint, the possible impact that the implementation of rules has on the representations of each operator. In this perspective, we can see if they have constructed what preventionists call the "spirit of safety." This reflects the expected level of security that the workers have for the company. The process requires that they establish a diagnosis of the company's safety level in reference to the regulations. This safety level is not always directly observable. In other words, an important part of their activity consists of converting the predicative form of highly prescriptive knowledge (regulations) into operative forms [36]. Aside from being expressed in the form of "rules of action," similar to "procedures," the knowledge should also contain concepts-in-action for the workers who will have to conceptualize it; this, however, is far from being easily achieved.

For example, preventionists set up a "spirit of safety" within their company in order to incite operators to understand the necessity of some formal rules. They choose their remarks according to what they know about their various interlocutors [37]. For example, cognitive-type reasoning that is related to the knowledge of hearing and its potential deterioration in certain industrial contexts is conveyed through concrete examples, displayed near the place where the hearing protection is kept.

Three of their meta-goals are identified:

1. Establishing a reference of what is expected from the formal interpretation of safety regulations, as well as what is estimated to be "a secure company."
2. Diagnosing the state of safety of the company: comparing every element of the "company system" in regard to the level of safety required by the regulations. This takes into account the context and knowledge of the company, for instance, in function of the all the machines and safety equipment (fire extinguishers, fire detectors, emergency exits, evacuation or workstation safety instructions, personal protection equipment, the sound level of an older machine, etc.), through constant verification of safety indicators, and by a priori and a posteriori analyses of incidents and/or accidents.
3. Upon establishing an understanding of formal references on one hand and diagnosing and cataloging the company's safety level on the other, preventionists have to reduce the distance between the two sides, by acting in a preventive or corrective way. They form plans of action or projects, trying to act before the accident or incident occurs, or else retroactively, via a process of "pragmatization of regulations."

In order to achieve each of these meta-objectives, different resources are mobilized. And we will assert that they form a system of instruments.

9.4.2 The System of Instruments of an Experienced Preventionist

First, we wish to recall the essential characteristics of a system of instruments as previously defined. A system of instruments is organized around a pivot instrument; it is finalized and oriented by the objectives of the worker's assignments; it fills various functions with redundancies and complementarities. A system of instruments is composed of heterogeneous artifacts and is also subjective and specific to a worker. We will illustrate each of these characteristics with different extracts from the operator's interview.

9.4.2.1 Regulations Are a Pivot Instrument in Finalized and Vicarious Systems

Subject A uses two different instruments: The Labor Code, his basic tool, and the Permanent Dictionary of Safety and Working Conditions. The latter is a version of the Labor Code that has been transformed. They simultaneously reach the same functions (redundancy), but they also have complementarity. The Permanent Dictionary gives additional information about regulations.

38. **Subject A**: Ah, okay, yes there is a **Labor Code, it is the fundamental tool**.
39. **Interviewer**: Because I don't put the documents ...
40. **Subject A**: the **Labor Code, the basic tool**, more than the Labor Code, for me what I use is the **Permanent Dictionary of Safety and Working Conditions**, because in fact it's the interpreted Labor Code. It is not simply the texts of laws, it is a little, it goes a little further, you don't have that in the Labor Code, and you have a certain number of orders, **decrees, and European directives**. So that allows you to simply have more information. That it is **the work tool**, yes, it is **the basic work tool**, it's true, I hadn't thought about that.

The third formal artifact concerns INRS[8] documents, referred to in the following comment:

50. **Subject A**: [...] it is also the advantage of the **much talked-about documentation of the INRS**, it's that it gives **more accessible information to everyone**, even if it goes into **technical aspects**, it starts by showing, if you take the case of a fire, they'll think of **explaining** what a fire is, you have **diverse levels of technicality** in this documentation.

These instruments are, first of all, useful for the elaboration of a diagnosis of the safety in work situations and for argumentation with the various actors of the company. All three instruments propose different levels of explanation and

---

[8] INRS: « Institut national de la recherche en sécurité » or French Research Institute of Safety.

detail. In order to implement safety regulations in the company, Subject A, for instance, can use the diversification of the technicality levels according the diversity of the actors. In doing so, he can "act on their representations" to increase "the spirit of safety" in his company [37].

To further this initial analysis, we have carried out a functional analysis of instruments. The instruments of the system elaborated by the operator fulfill multiple functions according to the three meta-goals previously presented, as well as the different sub-goal required to reach them. In the following table, we show an illustration of the instruments' functions identified in the interview. In the different columns, we indicate the considered artifacts to be used in action (potential instruments), their functions, the related meta-goals, and any comments (from extracts of the interview).

Table 9.2 presents an organized shape of the data. It allows to report functions supported by each of the artifacts on one hand and to consider the various corresponding meta-goals on the other. We can therefore see that:

1. Eleven functions are mobilized in their activity, supported by eleven artifacts. However, the distribution of the various functions is not homogeneous: on one hand certain artifacts support several functions and others only one. On the other hand, one identical function seems present in several artifacts, calling attention to the character of redundancy of functions known in systems of instruments. For example (cf. bold characters in the table, in the column "Functions"), we can note that the highly-rated function no. 2 (F2. to be capable of considering the risks of safety) is supported by 2 artifacts (Labor codes, rate of frequency and rate of gravity) or in the same way the highly-rated function no. 4 (F4. to simplify and "to contextualize" the elements of regulations) is supported by three artifacts (INRS documents, posters in workstations et informal communication).

2. When several artifacts and several functions fulfill the same meta-goal, there is evidence of the other character of systems of instruments: the complementarity of functions. For example (cf. bold characters in the table, the column "Meta-goals"), we can note that three artifacts (Labor code, Permanent Dictionary of Safety and Working Conditions and INRS documents) and five functions (F1–F5) jointly insure the highly rated meta-goal no. 1 (to Establish the expected reference of a "secure company").

Certain functions concern elements of the "pragmatization of regulations", by simplifying or contextualizing general rules or by completing some function (Table 9.2). For instance, concerning the first meta-objective, three types of artifacts are used complementarily: the Labor Code, the Permanent Dictionary and the INRS documents. Thus, informal communication concerns two meta-goals of the worker.

Table 9.2 Functional analysis of instruments related to the objectives of the preventionist

No.	Artifacts	Functions	Meta-goals	Comments
37...	Labor code	F1. To be capable of considering the statutory risks **F2. To be capable of considering the risks of safety**	**1. To Establish the expected reference of a "secure company"**	fundamental tool (38); basic tool (40); theoretical knowledge (62)
56	Permanent dictionary of safety and working conditions	F3. To update and widen the elements of the labor code	**1. To Establish the expected reference of a "secure company"**	
50	INRS documents	**F4. To simplify and "to contextualize" the elements of regulations** F5. Additionally, be capable of inferring production risks.	**1. To Establish the expected reference of a "secure company"**	On more technical aspects; more accessible
2–34	Frequency and gravity rate	**F2. To be capable of considering the risks of safety**	2. To Diagnose the current state of safety of the company	
42	Measuring instruments	F6. To measure the level of dust, the sound level, etc. F7. Be capable of measuring the risks of safety	2. To Diagnose the current state of safety of the company	
42	Grid of ergonomic analysis of workstation	F8. To estimate the risks and identify danger	2. To Diagnose the current state of safety of the company	
56	"All which is equipment"	F8. To estimate the risks and to identify danger	2. To Diagnose the current state of safety of the company	
62	Informal communication		2. To Diagnose the current state of safety of the company	
62	Posters in workstations	**F4. To simplify and "to contextualize" the elements of regulations** F9. To Be capable of deducing the risks	3. To Reduce the distance between theoretical reference and state of the real system, by prevention or correction, downstream either upstream to the accident or to the incident	Build the "spirit of safety" of people in the company

(continued)

Table 9.2 (continued)

No.	Artifacts	Functions	Meta-goals	Comments
62	Informal communication	**F4. To simplify and "to contextualize" the elements of regulations** F10- To be capable of deducing the risks	2. To Diagnose the current state of safety of the company 3. To Reduce the distance between theoretical reference and state of the real system, by prevention or correction, downstream either upstream to the accident or to the incident	Build the "spirit of safety" of people of the company
62	Causal tree	F11. To analyze a priori the accident or to the incident	3. To Reduce the distance between theoretical reference and state of the real system, by prevention or correction, downstream either upstream to the accident or to the incident	Build the "spirit of safety" of people of the company

9.4.2.2 A System of Instruments Compounding Heterogeneous Artifacts

In electrical maintenance, Vidal-Gomel [7, 8] considers rules as psychological instruments among other heterogeneous entities (symbolic or material), composing a coherent but hybrid system of instruments for the subject [4–6].

We have indeed identified different kinds of artifacts, which allow operators to diagnose the safety level in their company. These include measurement tools (for dust and sound levels), grids of observation (for workstation conformity), rates of frequency or severity, methods to understand the chain of causes of an accident,[9] and the reasoning to be maintained with the operators.

Concerning the rates of frequency and severity, Subject A explains that they are individual tools and collective ones (basic tools):

2. **Subject A**: The tools, the problem is because there aren't a lot, it is more like tools, **each person has their own tools**. For me, there are some basic tools, for example to calculate what we call **the rate of frequency, the rate of severity**, so that allows you, the rate of frequency allows you to determine the frequency of accidents, as its name indicates. And the rate of severity makes you determine the severity of the accidents. More or less.

---

[9] Interaction number 62 in the interview.

They are heterogeneous type resources: symbolic instruments (a method, a grid) or material ones (measuring instruments):

42. **Subject A**: Then after that, you have measuring tools, basically **measuring instruments**, which you can use, for example **to measure a sound level**, you are going to use tools, or more like instruments, **to measure noise, to measure the concentration of products, to measure the level of dust**, the rate of dust that you have in a workshop, [...] And then later you have especially **the more ergonomic aspect**, all of which is a **study of workstations**, you have grids, and you know maybe the R. grid?[10]

9.4.2.3 An Individualized System of Instruments

Concerning the preventionists' instruments, Subject A announces that each operator possesses his or her own, developed according to his or her individual experience and imagination. He also resorts to official indicators, each containing the same formula of calculation (for instance, for the rate of frequency and the rate of severity). During safety clubs, the professionals discuss their respective tools or the more informal uses of institutional or statutory tools.

9.5 Systems of Instruments, Training and Risk Management

Based on the instrumental approach, we consider that preventionists develop a system of instruments, of which we have found traces here.

We have utilized a comprehensive approach in analyzing a case study: the system of instruments of an experienced preventionist. More precisely, the inquiry is a theoretically oriented qualitative study [34], oriented by the investigation of a system of instruments. The first level of our analysis determines the components of the system; the second level is a functional analysis that allows us to identify the multi-functionalities of any artifacts used by the interviewed operator (identified in Table 9.2). Three artifacts fulfill the same function, for instance: "to establish the expected reference of a secure company" according to the Labor Code, the Permanent Dictionary or INRS Documents. So the actual system of the worker's instruments could authorize redundancies and vicarious characteristics, which contributes to a more resilient system. The adaptation permitted by this system of instruments imparts on more relevantly managed safety in the work situation.

These first results allow a developed discussion about the design of preventionists' training programs. It is astonishing to acknowledge that their use of safety

---

[10] It refers to an ergonomic grid of workstation observation developed in another firm.

rules is not taken into account during their training course [1]. The artifacts are exposed, but their functions and uses are not. In this way, novices are left to take their own initiative to understand each process during their field experience in companies, without prior conscience of the processes (e.g. of redundancies, for instance). They simply apply the rule, without considering the process of prag-matization. However, this is not specific to this particular training program. It is a general, frequently observed tendency. In safety training, safety rules and proce-dures are presented as the sole means of managing risks. Operators must simply apply them. This behaviorist approach is focused on technical dimensions and regulations; it denies both the cognitive dimensions of the activity through the use of rules, as well as the complexity of acting in a specific work organization [38]. From our point of view, a safety-training program must give the operators the means of understanding the different uses of safety rules and their safety functions, as well as the constraints of their uses in a context. Acting in a socio-technical system always involves arbitrating between different antagonistic goals (safety rules and time constraints, for example) [39]. If a training program ignores this, the operators will have to develop their own resources in each situation, with only their personal work experience, thus using the instruments with insufficient safety functions. To conclude, this is a very weak objective assigned to training.

A more ambitious objective would consist of taking the operators' systems of instruments into account. It would also involve guiding them during their training, to build effective resources for facing occupational risks and hazards in work situations. Training through periods of alternation seems a good means of reaching these objectives. After a period of practical experience, operators could give feedback during a formal training session. This would be a favorable moment to question what they have acquired through their own experience, and also to examine the system of instruments they have constituted, particularly in terms of functions, relationships (redundancy and complementarity) between the elements of the system and whether it has achieved sufficient safety functions. A functional analysis of the systems would allow them to determine any weaknesses or elements of robustness.

However, to go further, it would be necessary to pursue this analysis. The MFSR[11] [6] seems to be a good means to develop the operators' contributions to the resilience of sociotechnical systems. The use of MSFR could allow us to more systematically identify the functions and the reliability of the operators' system of instruments for a class of work situations. Then, by allowing the analysis of the fragility and the robustness of the operators' systems of instruments, MFSR could be considered a relevant tool to investigate the operators' contribution to the reliability of a work system. More generally, it would be useful to foster resilience in at-risk industrial systems. MFSR presents similarities with certain reliability methods, such as the *FMEA*,[12] in terms of structure and implementation in particular [4, 6].

---

[11] Method of Failure and Substitution of Resources.
[12] *FMEA*: Failure Mode and Effects Analysis, as a method of evaluation of the criticality of the consequences of the failures.

The resulting analysis would then not only be technical, as with FMEA, but focused on "anthropological" dimensions.

References

1. Munoz, G. (2007a). Champs conceptuels et champs d'activités chez les fonctionnels de sécurité. In M. Merri (Ed.), *Activité humaine et conceptualisation. Questions à Gérard Vergnaud* (pp. 527–538). Toulouse: Presses universitaires du Mirail.
2. Rabardel, P. (1995). *Les hommes et les technologies: approche cognitive des instruments contemporains.* Paris: Armand Colin.
3. Rabardel, P., & Bourmaud, G. (2003). From computer to instrument system: A developmental perspective. *Interacting with Computers, 15*(5), 665–691.
4. Bourmaud, G. (2006). *Les systèmes d'instruments : méthodes d'analyse et perspectives de conception.* Thèse de psychologie ergonomique. Saint-Denis: Université Paris 8 (http://buparis8.bu.inv-paris8.fr/web/collections/Page_theses_&_maitrises.php).
5. Bourmaud, G. (2007). L'organisation systémique des instruments : méthodes d'analyse, propriétés et perspectives de conception ouvertes. *Proceedings of ARCo'07: "Cognition, Complexité", Acta-Cognitica* (pp. 61–75). (http://hal.inria.fr/docs/00/19/11/28/PDF/061-076_Bourmaud.pdf).
6. Bourmaud, G. (2010). Proposition d'une méthode d'analyse anthropocentrée de la fiabilité et de l'adaptabilité des systèmes de travail. In A-S. Nyssen (Ed.), *SELF'2010, Ergonomics international congress: "Fiabilité, Adaptation et Résilience".* Liège, Belgique (http://www.ergonomie-self.org/media/media56015.pdf).
7. Vidal-Gomel, C. (2002a). Systèmes d'instruments des opérateurs. Un point de vue pour analyser le rapport aux règles de sécurité. *Pistes, 4*(2), (http://www.pistes.uqam.ca/v4n2/articles/v4n2a2.htm).
8. Vidal-Gomel, C. (2002b). Systèmes d'instruments: un cadre pour analyser le rapport aux règles de sécurité. In J.-M. Évesque, A.-M. Gautier, C. Revest, Y. Schwartz & J.-L. Vayssière (Eds.), *Proceedings du 37th Congrès de la SELF : Les évolutions de la prescription* (pp. 133–143). Aix-en-Provence: 25–27 septembre 2002, GREACT-SELF.
9. Munoz, G. (2007b). L'analyse de quelques « mouvements » entre les différentes formes de la connaissance : repères pour la formation. *Recherche en éducation, 4,* 39–50 (http://www.cren-nantes.net/spip.php?article72).
10. Viet, V., & Ruffat, M. (1999). *Les choix de la prévention.* Paris: Economica.
11. De la Garza, C., & Fadier, E. (2004). Sécurité et prévention: repères juridiques et ergonomiques. In P. Falzon (Ed.), *Ergonomie* (pp. 159–174). Paris: Presses universitaires de France.
12. Dekkers, S. (2006). Resilience engineering: chronicling the emergence of a confused consensus. In E. Hollnagel, D. D. Wood, & N. Leveson (Eds.), *Resilience engineering: Concepts and precepts* (pp. 77–92). Aldershot: Ashgate.
13. Hollangel, E. (2006). Resilience: the challenge of unstable. In E. Hollnagel, D., Woods & N. Leveson (Eds.), *Resilience engineering: Concepts and precepts* (pp. 9–17). Aldershot: Ashgate.
14. Daniellou, F. & Rabardel, P. (2005). Oriented approaches to ergonomics: sometraditions and communities. *Theoretical Issues in Ergonomics Science,* Vol. *6–5,* 353–357.
15. Faverge, J.-M. (1970). L'homme, agent d'infiabilité et de fiabilité du processus industriel. *Ergonomics, 13,* 301–327.

16. Munoz, G. & Bourmaud, G. (2007). Conceptualisation et pragmatisation de la réglementation comme instrument. In *proceedings of the Congrès of the Société Française de Psychologie 2007 and 4th Journées d'Etudes en Psychologie Ergonomique (EPIQUE'07)*. Nantes: 11–13 septembre 2007 (http://www.sfpsy.org/spe-grape/epique-2007/EPIQUE2007.pdf).
17. Leplat, J. (1998). About implementation of safety rules. *Safety Science, 29*, 189–204.
18. Hale, A. R., & Swuste, P. (1998). Safety rules: Procedural freedom or action constraint? *Safety Science, 29*, 163–177.
19. Reason, J. (1990). *Human error*. New York: Cambridge University Press.
20. Leplat, J. (2003). Questions autour de la notion de risque. In R. D. Kouabenan & M. Dubois (Eds.), *Les risques professionnels: évolutions des approches, nouvelles perspectives* (pp. 37–52). Toulouse: Octarès.
21. Hale, A. & Boris, D. (2012, in press). Working to rule or working safely ? Part 1: A state of the art review. *Safety science* (http://dx.doi.org/10.1016/j.ssci.2012.05.011).
22. Mayen, P., & Savoyant, A. (1999). Application de procédures et compétences. *Formation Emploi, 67*, 77–92.
23. Vergnaud, G. (1990). La théorie des champs conceptuels. *Recherches en didactique des mathématiques*, Vol. *10–2.3*, 133–170.
24. Faverge, J.-M. (1977). *Analyse de la Sécurité du Travail en Termes de Facteurs Potentiels d'Accidents*. Bruxelles: Université Libre de Bruxelles (Laboratoire de Psychologie Industrielle).
25. Folcher, V., & Rabardel P. (2004). Hommes-Artefacts-Activités : Perspective instrumentale. In P. Falzon (Ed.) *L'ergonomie* (pp. 251–268). Paris: Presses universitaires de France.
26. Bannon, L., & Bødker, S. (1991). Beyond the Interface, encountering artifacts in use. In J. Carroll (Ed.), *Designing interaction: Psychological theory of the human-computer interface* (pp. 227–253). New York: Cambridge University Press.
27. Kaptelinin, V., & Kuutti, K. (1999). Cognitive tools reconsidered: From augmentation to mediation. In J. Marsh, B. Gorayska & J. L. Mey (Eds.), *Humane interfaces: questions of methods and practice in cognitive technology* (pp. 145–160). Amsterdam, North-Holland: Elsevier.
28. Wertsch, J. V. (Ed.). (1985). *Culture, communication and cognition. Vygotskian perspectives*. New York: Cambridge University Press.
29. Lefort, B. (1982). L'emploi des outils au cours de tâches d'entretien et la loi de Zipf-Mandelbrot. *Le Travail Humain, 45*(2), 307–316.
30. Minguy, J.-L. (1995). *Concevoir pour aider à l'action située. Le travail en passerelle de navires de pêche : rôle de la carte de pêche comme représentation*. Thèse d'Ergonomie. Paris: Concervatoire national des arts et métiers.
31. Minguy, J.-L. (1997). Concevoir aussi dans le sillage de l'utilisateur. *International Journal of Design and Innovation Research, 10*, 59–78.
32. Zanarelli, C. (2003). *Caractérisation des stratégies instrumentales de gestion d'environnements dynamiques : Analyse de l'activité de régulation du métro*. Thèse de psychologie ergonomique. Saint-Denis: université Paris 8 (http://www.bu.univ-paris8.fr/web/collections/theses/zanarelli_catherine.pdf).
33. Vidal-Gomel, C., & Samurçay, R. (2002). Qualitative analysis of accidents and incidents to identify competencies. The electrical system maintenance case. *Safety Science, 40*(6), 479–500.
34. Creswell, J. W. (2007). *Qualitative inquiry and research design: Choosing among five approaches*. Thousand Oaks, CA: Sage.
35. Leplat, J. (1997). *Regard sur l'activité en situation de travail. Contribution à la psychologie ergonomique*. Paris: Presses universitaires de France.
36. Bruno, S., & Munoz, G. (2007). Développement et conversion en psychologie cognitive: possibilité d'une zone d'invariance minimale. In J. Baillé (Ed.), *Du mot au concept: Conversion* (pp. 47–73). Grenoble: Presses universitaires de Grenoble.

37. Munoz, G. & Bourmaud, G. (2011). De la conceptualisation des risques: le choix de l'argumentation des chargés de sécurité en fonction de leurs interlocuteurs. In the proceedings of *46th Congrès de la SELF: "L'ergonomie à la croisée des risques"* (Part 3, pp. 375–382). Paris: 14–16 septembre 2011 (http://www.ergonomie-self.org/media/media58466.pdf).
38. Mayen, P., & Vidal-Gomel, C. (2005). Conception, formation et développement de règles au travail. In P. Pastré & P. Rabardel (Eds.), *Modèles du sujet pour la conception, dialectiques activités-développement* (pp. 108–128). Toulouse: Octarès.
39. Leplat, J. (2008). *Repères pour l'analyse de l'activité en ergonomie*. Paris: Presses univeritaires de France.

Printed in the United States
By Bookmasters